LONDON MATHEMATICAL SOCIETY LECTURE NOTE SERIES

Managing Editor: Professor J.W.S. Cassels, Department of Pure Mathematics and Mathematical Statistics, University of Cambridge, 16 Mill Lane, Cambridge CB2 1SB, England

The books in the series listed below are available from booksellers, or, in case of difficulty, from Cambridge University Press.

London Mathematical Society Lecture Note Series. 169

Boolean Function Complexity

Edited by
M.S. Paterson
Department of Computer Science
University of Warwick

CAMBRIDGE
UNIVERSITY PRESS

Published by the Press Syndicate of the University of Cambridge
The Pitt Building, Trumpington Street, Cambridge CB2 1RP
40 West 20th Street, New York, NY 10011-4211, USA
10 Stamford Road, Oakleigh, Victoria 3166, Australia

First published 1992

Printed in Great Britain at the University Press, Cambridge

Library of Congress cataloguing in publication data available

British Library cataloguing in publication data available

ISBN   0 521 40826 1

Contents

Preface

Complexity theory attempts to understand and measure the intrinsic difficulty of computational tasks. The study of Boolean Function Complexity
reaches for the combinatorial origins of these difficulties. The field was pio-
neered in the 1950's by Shannon, Lupanov and others, and has developed now
into one of the most vigorous and challenging areas of theoretical computer
science.

In July 1990, the London Mathematical Society sponsored a Symposium
which brought to Durham University many of the leading researchers in the
subject for ten days of lectures and discussions. This played an important
part in stimulating new research directions since many of the participants
were meeting each other for the first time. This book contains a selection of
the work which was presented at the Symposium. The topics range broadly
over the field, representing some of the differing strands of Boolean Function
Theory.

I thank the authors for their efforts in preparing these papers, each of which
has been carefully refereed to journal standards. The referees provided in-
valuable assistance in achieving accuracy and clarity. Nearly all the referees'
names appear also in the list of authors, the others being A. Wigderson,
C. Sturtivant, A. Yao and W. McColl. While a measure of visual conformity
has been achieved (all but one of the papers is set using LaTeX), no attempt
was made to achieve uniform notation or a 'house style'. I have tried to
arrange the papers so that those which provide more introductory material
may serve to prepare the reader for some more austere papers which follow.
Some background in Boolean complexity is assumed for most of the papers.
A general introduction is offered by the three books by Dunne, Savage and
Wegener which are referenced in the first paper.

The Symposium at Durham was made possible by the initiative and sponsor-
ship of the London Mathematical Society, the industry and smooth organi-
zation of the staff at Durham University, the financial support of the Science
and Engineering Research Council and by the enthusiastic participation of the
Symposium members. Finally, I thank the staff and Syndics of Cambridge
University Press for their cooperation and patience during the preparation of
this volume.

Mike Paterson

University of Warwick
Coventry, England
June, 1992

Participants listed from left to right.

Top row: R. Raz, K. Edwards, N. Nisan, L. Valiant, A. Macintyre, K. Kalorkoti, W. Beynon, R. Smolensky, I. Newman, D. Uhlig, A. Chin, I. Leader, U. Zwick, C. Sturtivant, G. Brightwell.

Middle row: M. Jerrum, A. Cohen, A. Sinclair, A. Borodin, C. Schnorr, A. Wilkie, A. Andreev, N. Biggs, P. d'Aquino, M. Dyer, P. Dunne, A. Thomason.

Bottom row: I. Wegener, A. Stibbard, N. Pippenger, M. Klawe, J. Savage, M. Furst, A. Widgerson, W. McColl, M. Paterson, A. Yao, A. Razborov, M. Sipser, L. Henderson, C. McDiarmid, J. Shawe-Taylor, D. Barrington, R. Mirwald.

Participants in the
LMS symposium on Boolean Function Complexity,
Durham University, July 1990.

Miklos Ajtai
Alexander Andreev
David Mix Barrington
Alessandro Berarducci
Meurig Beynon
Norman Biggs
Béla Bollobás
Allan Borodin
Graham Brightwell
Andrew Chin
Aviad Cohen
Paola d'Aquino
Paul Dunne
Martin Dyer
Keith Edwards
Merrick Furst
Leslie Henderson

Mark Jerrum
Kyriakos Kalorkoti
Maria Klawe
Imre Leader
Angus Macintyre
William McColl
Colin McDiarmid
Roland Mirwald
Ilan Newman
Noam Nisan
Margarita Otero
Mike Paterson
Nicholas Pippenger
Franco Preparata
Ran Raz
Alexander Razborov

John Savage
Claus Schnorr
John Shawe-Taylor
Alistair Sinclair
Mike Sipser
Roman Smolensky
Alyson Stibbard
Carl Sturtivant
Andrew Thomason
Dietmar Uhlig
Leslie Valiant
Ingo Wegener
Avi Wigderson
Alex Wilkie
Andrew Yao
Uri Zwick

Relationships Between Monotone and Non-Monotone Network Complexity

Paul E. Dunne[*]

Abstract

Monotone networks have been the most widely studied class of restricted Boolean networks. It is now possible to prove superlinear (in fact exponential) lower bounds on the size of optimal monotone networks computing some naturally arising functions. There remains, however, the problem of obtaining similar results on the size of combinational (i.e. unrestricted) Boolean networks. One approach to solving this problem would be to look for circumstances in which large lower bounds on the complexity of monotone networks would provide corresponding bounds on the size of combinational networks.

In this paper we briefly review the current state of results on Boolean function complexity and examine the progress that has been made in relating monotone and combinational network complexity.

1. Introduction

One of the major problems in computational complexity theory is to develop techniques by which non-trivial lower bounds, on the amount of time needed to solve 'explicitly defined' decision problems, could be proved. By 'non-trivial' we mean bounds which are superlinear in the length of the input; and, since we may concentrate on functions with a binary input alphabet, the term 'explicitly defined' may be taken to mean functions for which the values on all inputs of length n can be enumerated in time 2^{cn} for some constant c.

[*] Department of Computer Science, University of Liverpool, Liverpool, L69 3BX, Great Britain.

Classical computational complexity theory measures 'time' as the number of moves made by a (multi-tape) deterministic Turing machine. Thus a decision problem, f, has *time complexity*, $T(n)$ if there is a Turing machine program that computes f and makes at most $T(n)$ moves on any input of length n.

The Turing machine is only one of many different models of computation. Another model, that has attracted as much attention, is the class of *combinational Boolean networks*. An *n-input combinational network* is a directed acyclic graph containing two distinct types of node: *input nodes*, which have no incoming edges; and *gate* nodes which have at most two incoming edges. Each input node is associated with a single Boolean variable, x_i, from an ordered set $\mathbf{X_n} = \langle x_1, x_2, \ldots, x_n \rangle$. Each gate node is associated with some two-input Boolean function. There is a unique gate, having no outgoing edges, which is called the *output* of the network. An assignment of Boolean values to the input variables naturally induces a Boolean value at the output gate, the actual value appearing depends on the input assignment and the network structure. The *size* of such a network is the number of gate nodes; its *depth* is the number of gates in the longest path from an input node to the output gate.

We shall denote by B_n the set of all n-input Boolean functions, $f(\mathbf{X_n})\{0,1\}^n \to \{0,1\}$ with formal arguments $\mathbf{X_n}$. An n-input combinational network computes $f \in B_n$ if for all assignments $\alpha \in \{0,1\}^n$ to $\mathbf{X_n}$, the value induced at the output gate is $f(\alpha)$. It should be noted that a single combinational network only solves a decision problem for the special case when all input strings are of length exactly n. In order to discuss the size (or *combinational complexity*) of networks for decision problems in general, the following approach is used. Let $[f_n]$ be the infinite sequence of Boolean functions arising by restricting a decision problem, f, to inputs of length n (thus $f_n \in B_n$). We say that the decision problem, f, is computed by a sequence of n-input combinational networks, $\langle C_n \rangle$, if, for each n, the n-input network, C_n, computes f_n. With this definition we can introduce appropriate complexity measures for Boolean functions computed by networks.

For a network, T, $\mathbf{C}(T)$ is the size of T; for a Boolean function $f \in B_n$

$$\mathbf{C}(f) = \min \{ \mathbf{C}(T) : T \ computes \ f \}$$

Finally for a family $[f_n]$ we say that the combinational complexity of $[f_n]$ is $g(n)$ if, for each f_n, it holds that $\mathbf{C}(f_n) \leq g(n)$. $\mathbf{D}(f)$ will denote the corresponding measure for depth.

If a decision problem can be computed in time $T(n)$ then $T(n)\log T(n)$ is an upper bound on the combinational complexity of the corresponding family of Boolean functions, see, e.g. Savage (1972), Schnorr

(1976a) or Fischer and Pippenger (1979). In this way sufficiently large lower bounds on combinational complexity would give similar bounds on Turing machine time. Lower bounds on Turing machine space could be obtained from $\omega(\log^2 n)$ lower bounds on combinational depth, cf. Borodin (1977).

In fact it is known that there are Boolean functions of n-arguments with exponential combinational complexity. Shannon (1949) proved that 'almost all'[1] $f \in B_n$ were such that $C(f) \geq 2^n/n$. Earlier, Riordan and Shannon (1942) had proved that, for almost all $f \in B_n$, $D(f) \geq n - \log \log n$. Lupanov (1958) (for size) and Gaskov (1978) (for depth) have established that these lower bounds are the best possible and so a lot is known about the difficulty of computing Boolean functions, by combinational networks, in the general case.

If we consider the case of explicitly defined Boolean functions, however, the existing results are extremely weak. To date, no superlinear lower bound has been proved on the combinational complexity of any specific function: the largest lower bound proved, is only $3n - 3$ for a function constructed in Blum (1984a). It has become clear that, if combinational networks are to provide a vehicle with which to derive superlinear lower bounds on Turing machine time — let alone resolve questions such as $P = ? NP$ — then techniques that are much more sophisticated, than those developed to date, must be constructed. In the absence of such methods, attention has been focused on restricted types of combinational networks. There are a number of reasons for proceeding along this path: one cannot hope to prove results on unrestricted networks unless one can prove results for special cases; understanding how to prove lower bounds on restricted types of network may give some insight into techniques that can be applied to the general case; and it may be possible to deduce lower bounds on combinational complexity from lower bounds on restricted networks, for example if the special class of networks can efficiently simulate combinational networks.

In this paper we are concerned with a particular class of restricted combinational network: monotone Boolean networks. These are introduced in Section 2, where a survey of lower bound results obtained for this model is also given. The remainder of the paper deals with the issue of relating monotone network complexity to combinational complexity: Section 3 describes a framework for translating between combinational and monotone networks and, within this, a class of functions known as *slice functions* may be shown to have closely related combinational and monotone network complexity. Slice functions and their properties are examined, in detail, in Section 4.

---

1) A property holds for 'almost all' $f \in B_n$ if the fraction of all n-input Boolean functions not possessing the property approaches zero as n approaches infinity.

Conclusions are given in the final section. The reader interested in progress on other aspects of combinational complexity or alternative restricted models may find discussions of work in these areas in Dunne (1988), Savage (1976), and Wegener (1987).

2. Monotone Boolean Networks

Combinational networks allow any two-input Boolean function to be used as a gate operation. The restriction imposed in the case of monotone Boolean networks is that the only gate operations admitted are two-input logical AND (or *conjunction*) — denoted $\wedge$ — and two-input logical OR (or *disjunction*) — denoted $\vee$. For Boolean variables x, y: $x \wedge y$ equals 1 if and only if both x and y equal 1; $x \vee y$ equals 1 if and only if at least one of x or y equals 1.

There is a penalty incurred by imposing this restriction on networks: it is no longer possible to compute every Boolean function of n arguments. In other words, the *basis* (i.e. permitted set of operations) $\{\wedge, \vee\}$ is *logically incomplete*. Post (1941) described necessary and sufficient conditions for a basis to be logically complete. In the next section we exploit two facts about complete bases, namely:

Fact 2.1: The basis $\{\wedge, \vee, \neg\}$ (where $\neg$ is the unary function corresponding to Boolean negation) is logically complete. $\square$

Fact 2.2: If $\Omega \subseteq B_2$ is a complete basis then the size of an optimal Boolean network, using only operations in Ω, computing a function $f \in B_n$ is at most $c\, C(f)$ for some (small) constant c. $\square$

A function which can be computed by a monotone Boolean network is called a *monotone Boolean function*. M_n denotes the (strict) subset of B_n comprising all n-input monotone Boolean functions. The study of this class of functions dates back to the work of Dedekind (1897) where the problem of calculating the exact value of $\psi(n) = |M_n|$ was first raised. This exact counting problem is still open, although asymptotically exact estimates have been obtained, cf. Korshunov (1981).

Monotone Boolean functions have a number of interesting properties which have proved important in constructing lower bound arguments for monotone network complexity. A few of these properties are summarised below.

Before stating these we need the following concepts. Define ordering relations $\leq$ and $<$ on Boolean functions as follows: $0 < 1$ and for f, g in B_n we say that $f \leq g$ if for all $\alpha \in \{0, 1\}^n$, $f(\alpha) = 1 \Rightarrow g(\alpha) = 1$. That is, whenever some assignment makes f take the value 1, the same assignment forces g to take the value 1. We say that $f < g$ if $f \leq g$ but f and g are

different functions. Now let f and g be functions in B_n with formal argu-
ments $\mathbf{X_n}$. $f^{|x_i := \varepsilon}$ denotes the function (in B_{n-1} with formal arguments
$\mathbf{X_n} - \{ x_i \}$) obtained by fixing x_i to the Boolean value ε.

Fact 2.3: Let $f \in B_n$ and let $\mathbf{X_n}$ be the formal arguments of f. $f \in M_n$ if
and only if: $\forall\, x_i,\ 1 \leq i \leq n$ it holds that $f^{|x_i := 0} \leq f^{|x_i := 1}$. $\square$

Fact 2.4: If f, g are in M_n and $f \leq g$ then:

i) $f \wedge g = f$

ii) $f \vee g = g$. $\square$

A conjunction of some subset of the variables $\mathbf{X_n}$ is called a *monom*. A
monom, m, is an *implicant* of $f \in M_n$ if $m \leq f$. A monom, m, is a *prime
implicant* of f if m is an implicant of f but no monom formed from a strict
subset of the variables of m is an implicant of f. $\mathbf{PI}(f)$ will denote the set
of prime implicants of f. The dual concepts, using disjunction, are clauses,
implicands, and prime clauses with $\mathbf{PC}(f)$ denoting the set of prime clauses
of a function f.

Fact 2.5: Any $f \in M_n$, with arguments $\mathbf{X_n}$, may be expressed uniquely in the
forms

$$f(\mathbf{X_n}) = \bigvee_{p \in \mathbf{PI}(f)} p \;\; ; \;\; f(\mathbf{X_n}) = \bigwedge_{q \in \mathbf{PC}(f)} q$$

The former is known as *Disjunctive Normal Form* (DNF); the latter as *Conjunctive Normal Form* (CNF). $\square$

$\mathbf{C^m}(f)$ will denote the monotone network complexity of $f \in M_n$ and $\mathbf{D^m}(f)$
the corresponding measure for monotone depth.

Early progress on the complexity of monotone Boolean networks was
similar to the case of combinational networks. Thus there are asymptotically
exact bounds for the monotone network size of almost all monotone Boolean
functions. The lower bound (of $2^n/n^{3/2}$) follows from Gilbert (1954) using
Shannon's arguments; the upper bound comes from Andreev (1988) (improving the constant factor in the construction of Red'kin (1979)).

The first significant development in the theory of monotone networks
came about with the appearance of superlinear lower bounds on the size of
monotone networks computing *sets* of monotone Boolean functions: 'superlinear' in this context means as a function of the total number of inputs and
outputs. Van Voorhis (1972) proved an asymptotically optimal lower bound
on the monotone network complexity of sorting n Boolean inputs; Paterson
(1975) and Mehlhorn and Galil (1976) independently obtained exact bounds
on the size of networks realising $(\wedge, \vee)$-Boolean matrix product; Weiss
(1984) and Blum (1984b) obtained lower bounds for the n-point Boolean
convolution function which is closely related to integer multiplication.

In the case of single monotone Boolean functions, until recently, as little progress had been made as for combinational networks. Although exact exponential lower bounds had been obtained by Schnorr (1976b) and Jerrum and Snir (1982) for monotone *arithmetic* networks (i.e. with only integer addition and multiplication permitted as operations) the techniques used to prove these results fail to work for algebraic structures in which the identities of Fact 2.4 hold. By the end of 1984 the most powerful techniques were capable of yielding only modest linear lower bounds, e.g. Dunne (1985), Tiekenheinrich (1984).

In 1985 the Soviet mathematician Razborov considered the following monotone Boolean functions.

Definition 2.1: Let $X_n^U = \{ x_{i,j} : 1 \leq i < j \leq n \}$ be a set of $N = n(n-1)/2$ Boolean variables representing the adjacency matrix of an *n*-vertex undirected graph $G(X_n^U)$. *k-clique* is the function in M_N, with formal arguments X_n^U, such that *k-clique*$(\alpha) = 1$ if the graph $G(\alpha)$ contains a *k*-clique, i.e. a set of *k* vertices every pair of which is joined by an edge of *G*.

Let $X_{n,n} = \{ x_{i,j} : 1 \leq i, j \leq n \}$ be a set of n^2 Boolean variables. The *Logical Permanent* is the function $PM \in M_{n^2}$, with formal arguments $X_{n,n}$, defined by

$$PM(X_{n,n}) = \bigvee_{\sigma \in S_n} \bigwedge_{i=1}^n x_{i,\sigma(i)}$$

where S_n is the set of all permutations of $\langle 1, 2, \ldots, n \rangle$. •

For appropriate (non-constant) values, the decision problem corresponding to the *k*-clique function is *NP*-complete.

Alon and Boppana (1986), improving the combinatorial arguments given originally in Razborov (1985a, 1985b), proved the following results concerning these functions.

Theorem 2.1: $\forall \; 3 \leq k < 0.25 \, (n/\log n)^{2/3}$

$$C^m(k-clique) \geq c \left(\frac{n}{16 \, k^{3/2} \log n} \right)^{\sqrt{k}} \qquad \square$$

Theorem 2.2:

$$C^m(PM) \geq n^{c \log n} \qquad (\forall \; c < 1/16) \qquad \square$$

The lower bound of Theorem 2.1 is exponential for large enough values of *k*. In addition to these results of Razborov, Alon, and Boppana, exponential lower bounds on explicitly defined monotone Boolean functions have been proved in Andreev (1985, 1987) and Tardos (1987).

Theorems 2.1 and 2.2 constitute a significant advance in the theory of Boolean network complexity since they are built on a technique which is powerful enough to yield superlinear lower bounds on size for a non-trivial network model. Further indications that monotone networks are a theoretically tractable model are given by the methods of Karchmer and Wigderson (1987) and Raz and Wigderson (1990). Their results concern the depth of monotone networks.

Definition 2.2: The function $st{-}conn(\mathbf{X}_n^U)$ is the monotone Boolean function such that $st{-}conn(\alpha) = 1$ if $G(\alpha)$ contains a path from vertex s to vertex t.

Theorem 2.3: (Karchmer and Wigderson, 1987)

$$\mathbf{D}^m(st - conn) = \Omega(\log^2 n) \qquad \square$$

Theorem 2.4: (Raz and Wigderson, 1990)

$$\mathbf{D}^m(PM) = \Omega(n) \qquad \square$$

Razborov (1988) also proves superlogarithmic lower bounds on monotone depth.

3. A Framework for Relating Combinational and Monotone Network Complexity

The theorems stated at the conclusion of the preceding section may be regarded as completing the first part of a programme aimed at achieving non-trivial lower bounds on problem complexity. Thus, for the restricted case of monotone networks, techniques powerful enough to prove large lower bounds on size and depth are known. The question that now arises is: how relevant are these results/techniques to combinational complexity? In other words: is it possible to deduce non-trivial lower bounds on combinational complexity (depth) from large enough lower bounds on monotone complexity (depth)?

The results of Razborov (1985b), Tardos (1987) and Raz and Wigderson (1990), at first sight, offer a negative answer to the second question.

Theorem 3.1:

i) $C(PM) = O(n^k)$ for some constant k.

ii)   There is function computable with polynomial size combinational networks that requires exponential size monotone networks.

iii) There is a function computable in $O(\log n)$ depth using combinational networks that requires $\Omega(\sqrt{n})$ depth monotone networks.

Proof: (i) follows by observing that the Logical Permanent is equivalent to determining whether a given bipartite graph contains a perfect matching. Hopcroft and Karp (1973) give a polynomial time algorithm for this problem and thus the upper bound on combinational complexity is immediate. (ii) is proved in Tardos (1988) and (iii) by Raz and Wigderson (1990). □

The second and third parts of Theorem 3.1 (which are both proved using explicitly defined functions) show that there are exponential gaps between monotone network size (depth) and combinational network size (depth). As a consequence it will not *always* be possible to derive lower bounds on combinational complexity using lower bounds on monotone complexity. Nevertheless the theorem does not exclude the possibility of doing this for *some* monotone Boolean functions.

Recall from Facts 2.1 and 2.2 that the basis $\{\wedge, \vee, \neg\}$ is logically complete and that an optimal Boolean network built from any complete basis of two-input Boolean operations is at most a constant factor larger than an equivalent optimal combinational network. It follows that, since we are interested in superlinear lower bounds, we may without loss of generality consider the problem of relating monotone networks to networks which only permit the operations $\{\wedge, \vee, \neg\}$ to be used.

$\{\wedge, \vee, \neg\}$-networks only differ from monotone networks in permitting the use of negation. The result below demonstrates that we can make such networks more closely resemble monotone networks by permitting the use of negation only on *input* nodes. We shall use $C_{\{\wedge, \vee, \neg\}}(f)$ to denote the number of gates in the smallest $\{\wedge, \vee, \neg\}$-network realising $f \in B_n$.

Definition 3.1: A *standard network* is a Boolean network whose permitted gate operations are $\{\wedge, \vee\}$ and with $2n$-input nodes:

$$\langle x_1, \ldots, x_n, \neg x_1, \ldots, \neg x_n \rangle$$

$SC(f)$ will denote the number of *gate* nodes in the smallest standard network realising $f \in B_n$. •

Theorem 3.1: $\forall f \in B_n$ it holds that $SC(f) \leq 2 C_{\{\wedge, \vee, \neg\}}(f)$.

Proof: (Outline) The following identities (known as De Morgan's Laws) can be easily proved:

$$\neg(x \wedge y) = (\neg x) \vee (\neg y) \quad ; \quad \neg(x \vee y) = (\neg x) \wedge (\neg y)$$

Let T be an optimal $\{\wedge, \vee, \neg\}$-network realising some $f \in B_n$. Let g be a 'last' gate in T such that an edge directed out of g enters a negation gate. Here 'last' means that no gate on a path from g to the output gate has the property that an edge directed out of it enters a negation gate. Now since we include instances of negation in measuring size and we have assumed that T is optimal it follows that there is *exactly one* wire leaving g and entering a negation gate, h say. Let $h_1, \ldots, h_r$ be the gates which have h as an input. Let g_1 and g_2 be the gates supplying the inputs of g. We change T as follows: add a new gate g' whose inputs are $\neg g_1$ and $\neg g_2$; remove the negation gate h and replace each edge $\langle h, h_i \rangle$ by an edge $\langle g', h_i \rangle$; finally if g is an $\land$-gate then make g' an $\lor$-gate and vice versa. From De Morgan's Laws it follows that the new network, T', still computes f.

Applying the process of the preceding paragraph repeatedly, we eventually reach the situation where only input nodes enter a negation gate. Since we add only one new ($\land$ or $\lor$) gate at each stage it follows that the final network is a standard network computing f and containing at most twice the number of gates in T. $\square$

Now consider an optimal combinational network, T, computing some $f \in M_n$. This may be transformed to a standard network, S, that also computes f, and is only a constant factor larger than T. The only way in which S differs from a monotone network is by the presence of the n extra input nodes $\langle \neg x_1, \ldots, \neg x_n \rangle$.

Suppose that we, temporarily, ignore the fact that the n additional inputs are the negation of the n function arguments and regard them as n new Boolean variables $y_1, \ldots, y_n$. Then it is clear that:

i) S computes a *monotone* Boolean function of the inputs $\langle x_1, \ldots, x_n, y_1, \ldots, y_n \rangle$.

ii) If, for each i, we substitute $\neg x_i$ for the input y_i then S computes the original function $f \in M_n$.

One of the most important techniques applied in proving lower bounds on monotone network complexity is the concept of *replacement rules*. These prescribe 'circumstances' in which a node of a monotone network computing some function $h(\mathbf{X_n})$ may be *replaced* by a node computing some *different* function $h'(\mathbf{X_n})$ without altering the function, f, computed by the network. The 'circumstances' depend solely on h, h' and f and *not* on the topology of the network.[2]

Returning to the standard network S in which $\neg x_i$ is regarded as a new input y_i we can attempt to use the concept of replacement rules to yield a monotone network with inputs $\mathbf{X_n}$ which computes f. Thus, if the following two conditions can be satisfied, *for all* standard networks computing f, we may deduce that $\mathbf{C^m}(f)$ is 'not much larger' than $\mathbf{C}(f)$.

C1) There is a set $\langle h_1, \ldots, h_n \rangle$ of monotone Boolean functions having formal arguments $\mathbf{X_n}$ such that replacing any subset of the y_i inputs by the

---

2) The power of this technique arises from the fact that one may identify functions which can be replaced by the Boolean constants 0 or 1 and thus cannot be computed as partial results in optimal monotone networks. An example of the technique in practice may be found in Paterson (1975). A full characterisation of applicable replacements is given in Dunne (1984, 1988), see also Beynon's paper in this volume.

corresponding h_i functions and the remaining y_j inputs by the corresponding $\neg x_j$ inputs, results in a network computing f.

C2) The set of n monotone Boolean functions $\langle h_1, \ldots, h_n \rangle$ can be computed by a *monotone* network of size at most $\varepsilon_n \, \mathbf{C^m}(f)$ (for some $\varepsilon_n < 1$).

Theorem 3.3: If $f \in M_n$ for which conditions (C1) and (C2) hold, then

$$\mathbf{C}(f) \geq \frac{1 - \varepsilon_n}{2c} \mathbf{C^m}(f)$$

where c is the constant of Fact 2.2.

Proof: If both (C1) and (C2) hold then it follows that $\mathbf{C^m}(f) \leq \mathbf{SC}(f) + \varepsilon_n \mathbf{C^m}(f)$. The theorem now follows from Fact 2.2 and Theorem 3.2. □

For $f \in M_n$, a set $\langle h_1, \ldots, h_n \rangle$ of monotone functions satisfying condition (C1) for f, is called a *pseudo-complement vector for f*. h_i is called a *pseudo-complement* for x_i when computing f. Informally a pseudo-complement for x_i can replace the node $\neg x_i$ in *any* standard network computing f.

Given the relation in Theorem 3.3, it is clearly desirable to identify classes of monotone Boolean functions for which both conditions (C1) and (C2) hold. In fact it turns out that (C1) holds *for all $f \in M_n$*.

Theorem 3.4: $h \in M_{n-1}$ with formal arguments $\mathbf{X_n} - \{x_i\}$ is a pseudo-complement for x_i when computing $f \in M_n$ (with arguments $\mathbf{X_n}$) if and only if

$$f^{|x_i := 0} \leq h \leq f^{|x_i := 1}$$

Proof: The result was originally proved in Dunne (1984). This proof is reproduced in Dunne (1988) pp. 242-243. □

Corollary 3.1: $\forall \; f \in M_n$ condition (C1) holds.

Proof: From Fact 2.3, $f \in M_n$ if and only if $f^{|x_i := 0} \leq f^{|x_i := 1}$ for each x_i. It follows that the interval of Theorem 3.4 is always well-defined. □

Theorem 3.4 does not, however, allow functions for which condition (C2) holds to be identified directly. An 'obvious' choice of pseudo-complement vector, such as the n subfunctions of f obtained by fixing x_i to 0, will not give an efficient transformation from standard networks to monotone networks. Theorem 3.4 is mainly of use in permitting simple proofs of the *correctness* of specific pseudo-complements.

Rather than attempt to identify, explicitly, those $f \in M_n$ for which (C2) holds, i.e. for which efficiently computable pseudo-complement vectors exist, we proceed in the 'reverse direction'. Thus:

Suppose we are given a function $g \in M_n$ which is to be used as a pseudo-complement (by constructing some simple variant, h, of g). For which $f \in M_n$ is it possible to combine g and f into some new function $F(f, g)(\mathbf{X_n})$ such that

$\langle h^{\lfloor x_1 := 1}, \ldots, h^{\lfloor x_n := 1} \rangle$ is a pseudo-complement vector for $F(f, g)$?

Finding a suitable function $F(f, g)$ would allow *any* monotone Boolean function, g, to be used as a mechanism for constructing efficient pseudo-complements. Furthermore, large enough lower bounds on the monotone complexity of $F(f, g)$ would give superlinear lower bounds on the *combinational* complexity of f.

The next theorem describes a uniform method of constructing a suitable $F(f, g)$ which applies to *any* $f \in M_n$ and *any* $g \in M_n$.

Theorem 3.5: Let $f, g \in M_n$ with arguments $\mathbf{X_n}$. Define the function h by $h = \bigvee_{i=1}^{n} (x_i \wedge g^{\lfloor x_i := 0})$ and the function $F(f, g)$ by $F(f, g) = (f \wedge g) \vee h$.

$h^{\lfloor x_i := 1}$ is a pseudo-complement for x_i when computing $F(f, g)$.

Proof: From Theorem 3.4 it suffices to prove that

$$F^{\lfloor x_i := 0} \leq h^{\lfloor x_i := 1} \leq F^{\lfloor x_i := 1}$$

$$
\begin{aligned}
F^{\lfloor x_i := 0} &= (f^{\lfloor x_i := 0} \wedge g^{\lfloor x_i := 0}) \vee h^{\lfloor x_i := 0} \\
&\leq g^{\lfloor x_i := 0} \vee h^{\lfloor x_i := 0} \\
&\leq g^{\lfloor x_i := 0} \vee h^{\lfloor x_i := 1} && \text{(monotonicity)} \\
&= h^{\lfloor x_i := 1} && \text{(Fact 2.4 and choice of } h) \\
&\leq (f^{\lfloor x_i := 1} \wedge g^{\lfloor x_i := 1}) \vee h^{\lfloor x_i := 1} \\
&= F^{\lfloor x_i := 1} && \square
\end{aligned}
$$

Corollary 3.2: For all $f \in M_n$, $g \in M_n$ if $F(f, g)$ and h are defined as in Theorem 3.5 then:

$$\mathbf{C^m}(F(f, g)) \leq O(\mathbf{C}(F(f, g))) + n^2 \, \mathbf{C^m}(g) \tag{i}$$

$$\mathbf{C}(F(f, g)) \leq \mathbf{C}(f) + (n + 1)\mathbf{C}(g) + 1 \tag{ii}$$

$$\mathbf{C}(f) = \Omega(\mathbf{C^m}(F(f, g)) - (n^2 + n - 1)\mathbf{C^m}(g)) \tag{iii}$$

Proof: (i) follows from Theorem 3.2 and Theorem 3.4 since the n pseudo-complements can be computed using at most $n\,\mathbf{C^m}(h) \leq n^2 \, \mathbf{C^m}(g)$ monotone gates. (ii) follows from the definition of F and the fact that $\mathbf{C}(h) \leq n\,\mathbf{C}(g) - 1$. (iii) is an immediate consequence of (i) and (ii). $\square$

It should be noted that the upper bounds, on the monotone complexities of h and the pseudo-complement vector, used in the corollary are extremely crude. For specific g better estimates are possible. Nevertheless the corollary shows that if we have a monotone function g with linear monotone complexity, combine this with some monotone function f, as described, and can then prove a lower bound of $\omega(n^3)$ on the monotone complexity of the resulting $F(f,g)$, then we can deduce that the *combinational* complexity of f is also $\omega(n^3)$.

In summary the programme to prove non-trivial lower bounds on combinational complexity has advanced to a stage where one should investigate the behaviour of specific monotone functions g in the context of Theorem 3.5 and Corollary 3.2.

There is a particular class of monotone Boolean functions that has been extensively studied: the *threshold functions*. The k-th threshold function of n arguments — denoted T_k^n — is the function defined to be 1 whenever at least k of its arguments take the value 1.

Fact 3.1:

i) $C^m(T_k^n) = O(n)$ for k or $n-k$ constant.

ii) $C^m(T_k^n) = O(n \log n)$ for $k = \omega(1)$.

iii) $C(T_k^n) = O(n)$ for all k. □

Suppose g in the statement of Theorem 3.5 is chosen to be T_k^n. We wish to determine the precise forms that this leads to for the functions h and $F(f,T_k^n)$. Both of these functions turn out to have a particularly simple form.

Fact 3.2: If $g = T_k^n$ in Theorem 3.5 then

$$h = \bigvee_{i=1}^{n} (x_i \wedge g^{|x_i:=0}) = T_{k+1}^n \qquad\qquad □$$

It follows from Fact 3.2 that the function $F(f,T_k^n)$ is of the form

$$(f \wedge T_k^n) \vee T_{k+1}^n$$

The function $F(f,T_k^n)$ is called the *k-slice* of f and will be denoted $f^{(k)}$. It should be noted that, since there are n (non-constant) threshold functions on n variables, every $f \in M_n$ has n distinct slice functions associated with it. In fact it will be convenient to consider an additional slice function $f^{(0)}$, which is defined as $f \vee T_1^n$, cf. Theorem 4.1. From Theorem 3.5 and Fact 3.2 we immediately have

Fact 3.3: $\forall f \in M_n$, $\forall 1 \le k \le n$: $T_k^{n-1}(\mathbf{X_n} - \{x_i\})$ is a pseudo-complement for x_i when computing the k-slice of f. □

Historical note: It must be pointed out that we have (deliberately) distorted the chronological order of the results presented in this section for tutorial purposes. The discovery of slice functions and the fact that these have closely related monotone and combinational complexity, was made by Berkowitz (1982). Thus Fact 3.3 was first proved by Berkowitz outside the framework developed above. Theorem 3.4 appears in Dunne (1984) as a generalisation of Berkowitz' transformation for slice functions. The results stated as Theorem 3.5 and Corollary 3.2 are recent work of the author.

4. Slice Functions and their Properties

If we examine the structure of a k-slice function, $f^{(k)}$, more closely, the values returned by it are seen to have a simple characterisation in terms of the set of possible input assignments. Thus: for any assignments α in which fewer than k inputs take the value 1, $f^{(k)}(\alpha) = 0$; for any assignments, α in which more than k inputs are 1, $f^{(k)}(\alpha) = 1$; and for any assignments, α, in which exactly k inputs are 1, $f^{(k)}(\alpha) = f(\alpha)$.[3] One important consequence of this behaviour is that $f^{(k)}$ is always a monotone Boolean function, even if f is non-monotone. In this way it is, in principle, possible to obtain lower bounds on the combinational complexity of non-monotone functions from lower bounds on the monotone complexity of related functions.

We observed after Corollary 3.2 that the upper bounds in that result could be improved for specific choices of g. Slice functions provide an example of this. In addition they have a property which is not guaranteed by Corollary 3.2, namely: if the combinational complexity of f is large enough then some slice function of f must have superlinear combinational complexity. In more precise terms the important relationships between monotone and combinational complexity for slice functions are summarised by the theorem below.

Theorem 4.1: For all $f \in B_n$:

i) $\quad C(f) \leq \sum_{k=0}^{n} C(f^{(k)}) + O(n)$

ii) $\quad C(f^{(k)}) \leq C(f) + O(n)$

iii) $\quad C^m(f^{(k)}) = O(C(f^{(k)}) + s(n, k))$ where $s(n, k) = n$ for k or $n-k$ constant and $n \log^2 n$ otherwise.

---

3) Of course, a similar behavioural analysis can be made for any function constructed along the lines of Theorem 3.5. In practice this turns out to be less elegant than the corresponding analysis for slice functions.

Proof: For (i) let E_k^n be the (non-monotone) Boolean function which is 1 if exactly k inputs are 1. Then any $f \in B_n$ may be expressed as $\bigvee_{i=0}^{n} (f \wedge E_k^n)$. (i) now follows from Fact 3.1(iii) and the identity $f \wedge E_k^n = f^{(k)} \wedge (\neg T_{k+1}^n)$. (ii) is from Fact 3.1(iii). (iii) is a consequence of Theorem 3.2 using the result that the pseudo-complement vector

$$\langle T_k^{n-1}(\mathbf{X_n} - \{ x_1 \}), \ldots, T_k^{n-1}(\mathbf{X_n} - \{ x_n \}) \rangle$$

can be computed using only $O(s(n,k))$ monotone gates. These upper bounds were obtained, independently, by Paterson (personal communication, 1984); McColl (personal communication, 1984 for the case k constant); Wegener (1985) and Valiant (1986). □

Combining the inequalities (i)-(iii) we see that $f \in B_n$ has 'large' combinational complexity *if and only if* some $f^{(k)}$ has 'large' monotone complexity. i.e.

$$\mathbf{C^m}(f^{(k)}) = \omega(s(n,k)) \quad \Rightarrow \mathbf{C}(f^{(k)}) = \omega(s(n,k)) \qquad \text{(From iii)}$$

$$\Rightarrow \mathbf{C}(f) = \omega(s(n,k)) \qquad \text{(From ii)}$$

In the opposite direction,

$$\mathbf{C}(f) = \omega(n^2 \log^2 n) \Rightarrow \exists k \ s.t. \ \mathbf{C}(f^{(k)}) = \omega(n \log^2 n) \qquad \text{(From i)}$$

$$\Rightarrow \exists k \ s.t. \ \mathbf{C^m}(f^{(k)}) = \omega(n \log^2 n) \qquad \text{(From iii)}$$

If f is a Boolean function with superpolynomial combinational complexity then some k-slice function of f must have superpolynomial monotone complexity. If we could identify a *specific* value of k for which $f^{(k)}$ could be *proved* to have superpolynomial monotone complexity then this proof would give a superpolynomial lower bound on the combinational complexity of f and, hence, a similar bound on the time needed to solve the decision problem corresponding to f.

Undoubtedly the most extensively studied class of decision problems that are believed to have superpolynomial time complexity is the class of *NP*-complete problems. These were introduced by Cook (1971). For our purposes we merely observe that this class contains many important combinatorial and optimisation problems for which no efficient (i.e. polynomial time) algorithms are known. It is believed that no such algorithms exist and a proof that a single *NP*-complete problem required superpolynomial time would immediately establish that all *NP*-complete problems need superpolynomial time algorithms. An excellent survey of work on this class of decision problems is given by Garey and Johnson (1979).

From the development above we could prove such a bound by obtaining large enough bounds on the combinational complexity of the Boolean function family corresponding to some NP-complete decision problem, i.e. by proving superpolynomial bounds on the monotone complexity of some slice function of an NP-complete Boolean function. This fact motivates the following question: for a given NP-complete Boolean function, f, which slices of f 'are likely' to be difficult?

The k-clique function was introduced in Section 2. For $k = n/2$ the decision problem corresponding to $(n/2) - clique$ is NP-complete. To simplify the notation we shall denote this function by HCL_n (for Half Clique). Recall that this has $N = n(n - 1)/2$ formal arguments $\mathbf{X_n^U} = \{ x_{i,j} : 1 \le i < j \le n \}$ encoding the edges of an n-vertex undirected graph $G(\mathbf{X_n^U})$.

Any prime implicant of HCL_n corresponds to some potential $(n/2)$-clique in $G(\mathbf{X_n^U})$. So every prime implicant of this function contains exactly $(n/4)(n/2 - 1)$ variables (corresponding to the edges in the particular clique). Any $f \in M_n$ with the property that each prime implicant of f contains exactly m variables has a special slice function, called the *canonical slice*, denoted $c - sl(f)$: this is just the slice function $f^{(m)}$. For the canonical slice we have

$$c - sl(f) \ = \ f^{(m)} \ = \ f \vee T_{m+1}^n$$

$HCL_n(\mathbf{X_n^U})$ ought to have a slice with superpolynomial complexity. The result below summarises which slices may be eliminated from consideration.

Theorem 4.2: Let $b(n) = (n/4)(n/2 - 1)$ and $t(n) = n(n - 2)/2 - 3$. The table below gives upper bounds for $\mathbf{C^m}(HCL_n^{(k)})$ for specific values of k.

	k	$\mathbf{C^m}(HCL_n^{(k)})$
i)	$< b(n)$	$O(N \log N)$
ii)	$b(n)$	$O(N \log N)$
iii)	$b(n) + c$	$O(N^{c+1} \log N)$
iv)	$t(n)$	$O(N^{3/2})$
v)	$> t(n)$	$O(N \log N)$

Proof: The upper bounds in (i) and (v) are similar: any graph with fewer than $b(n)$ edges cannot contain an $n/2$-clique so $HCL_n^{(k)}$ in these cases is just the threshold function T_{k+1}^N and the upper bound follows from Fact 3.1(ii). Analogously, any graph with more than $t(n)$ edges is guaranteed to contain an $(n/2)$-clique by Turan's Theorem (cf. Bollobás (1978) pp. 292-295), so in these cases $HCL_n^{(k)}$ is the threshold function T_k^N.

The upper bound in (ii) was proved in Wegener (1985). Consider

$$HCL_n^{(b(n))} \ = \ (HCL_n \wedge T_{b(n)}^N) \vee T_{b(n)+1}^N$$

An equivalent function to this is obtained by replacing HCL_n with *any* function, g, such that $g(\alpha) = HCL_n(\alpha)$ for all assignments α having exactly $b(n)$ variables set to 1, cf. the remarks at the opening of this section. It is easy to show that an n-vertex graph with exactly $b(n)$ edges contains an $(n/2)$-clique if and only if at least $n/2$ of the vertices have degree at least $(n/2) - 1$. Letting $X^{(i)}$ denote the variables

$$X^{(i)} = \{ x_{j,i} : 1 \le j < i \} \cup \{ x_{i,j} : i < j \le n \}$$

It follows that this condition is captured by the function

$$g(\mathbf{X_n^U}) = T_{n/2}^n (T_{n/2-1}^{n-1}(X^{(1)}), \ldots, T_{n/2-1}^{n-1}(X^{(n)}))$$

$\mathbf{C^m}(g) = O(N \log N)$ and from the previous argument g may replace HCL_n in $HCL_n^{(b(n))}$.

The upper bound in (iii) is proved in Dunne (1986). Finally (iv) is similar to (ii) using the fact that Turan's Theorem characterises those graphs with $t(n)$ edges which do not contain an $(n/2)$-clique. □

Results similar to Theorem 4.2 can also be proved for Hamiltonian Circuit (both directed and undirected) and for a suitable encoding of Satisfiability, cf. Dunne (1986). For all of these cases the functions are monotone and have a canonical slice.

Theorem 4.2 demonstrates that if k is too small or too large, in defining $HCL_n^{(k)}$, then the slice function is easy to compute. In order to identify a specific slice function of HCL_n which is likely to be difficult we choose a value of k between these two extremes.

Definition 4.1: Let $f \in B_n$. The *central slice* of f (denoted $Cen(f)$) is the $\lceil n/2 \rceil$-slice of f. •

Definition 4.2: Let $f \in B_n$ with formal arguments $\mathbf{X_n}$ and $g \in B_p$ $(p \ge n)$ with formal arguments $\mathbf{Y_p} = \langle y_1, \ldots, y_p \rangle$. f is a *projection* of g if there is a mapping

$$\sigma : \mathbf{Y_p} \to \mathbf{X_n} \cup \{ \neg x_1, \ldots, \neg x_n \} \cup \{ 0, 1 \}$$

such that $f(\mathbf{X_n}) \equiv g(\sigma(\mathbf{Y_p}))$. f is a *monotone projection* of g if the mapping σ does not contain negated variables in its image. •

Theorem 4.3: (Dunne, 1986) The decision problem corresponding to $Cen(HCL_n)$ is *NP*-complete.

Proof: We show that the $(n/2)$-clique problem is polynomially reducible to the decision problem corresponding to $Cen(HCL_n)$. The proof works by constructing (uniformly) a non-monotone projection from $Cen(HCL_{5n})$ onto $(n/2)$-clique.

We proceed as follows. Given an n-vertex graph $G(V, E)$ we build a $5n$-vertex graph $H(W, F)$ with the properties:

i) H contains exactly $(5n/4)(5n - 1)$ edges.

ii) H contains a $(5n/2)$-clique if and only if G contains an $(n/2)$-clique.

H consists of 3 parts: a copy of the graph G; a copy of the *complement graph* $\bar{G}$ (i.e. the graph which contains an edge between i and j if and only if G does *not* contain such an edge); and a $3n$-vertex graph Q which satisfies the following conditions:

Q1) Q contains a $(2n)$-clique $\langle q_1, \ldots, q_{2n} \rangle$.

Q2) Q does not contain a $(2n + 1)$-clique.

Q3) Q contain exactly $(7n^2 + n)/4$ edges other than those in the $(2n)$-clique.

In addition there are edges from every vertex of G to every vertex q_i in the $(2n)$-clique of Q. H contains no other edges. In terms of the projection, σ, which this represents, exactly one formal argument of $Cen(HCL_{5n})$ will be mapped to $x_{i,j}$ in $\mathbf{X}_n^U$ and exactly one to $\neg x_{i,j}$. All the remaining arguments will be fixed to constants in accordance with the definition of H. Note that Q exists (and is easily constructible) as a consequence of Turan's Theorem.

In total, H will contain:

$$|G| + |\bar{G}| + n(2n - 1) + (7n^2 + n)/4 + 2n^2$$

edges. Since $\bar{G}$ is the complement of G it follows that $|G| + |\bar{G}|$ is exactly $n(n - 1)/2$. Thus H contains exactly $(5n/4)(5n - 1)$ edges, as required.

It remains to show that G contains an $(n/2)$-clique if and only if H contains a $(5n/2)$-clique. Suppose $\{v_1, \ldots, v_{n/2}\}$ is an $(n/2)$-clique in G. Since every vertex of G is joined to each vertex q_i in the $(2n)$-clique of Q this means that

$$\langle v_1, \ldots, v_{n/2}, q_1, \ldots, q_{2n} \rangle$$

is a $(5n/2)$-clique in H. On the other hand, suppose that H has a $(5n/2)$-clique. This cannot use any vertices in $\bar{G}$, since $\bar{G}$ has only n vertices and is disconnected from the rest of H. In addition this clique can use at most $2n$ vertices of Q since Q does not contain a $(2n + 1)$-clique. It follows that any such clique must contain at least $n/2$ vertices from G and thus G contains an $(n/2)$-clique.

This completes the proof that $Cen(HCL_n)$ is NP-complete. $\square$

Corollary 4.1: $\mathbf{C}(HCL_n) = O(\mathbf{C}^{\mathbf{m}}(Cen(HCL_{5n})))$. $\square$

So if $Cen(HCL_n)$ has small monotone complexity then every NP-complete Boolean function has polynomial combinational complexity.

Dunne (1986) proves similar results for the central slice functions of the Hamiltonian circuit functions and the central slice of Satisfiability. It is not the case, however, that the central slice function of every NP-complete Boolean function is NP-complete. A counterexample is given by the Cubic Subgraph function (see Garey and Johnson (1978), p. 198). Erdös and Simonovits (1973) prove that every n-vertex graph containing $\Omega(n^{5/3})$ edges contains a cubic subgraph.

5. Conclusion

In this paper we have considered the problem of translating lower bounds on the complexity of monotone networks to lower bounds on combinational complexity. Despite the fact that it will not always be possible to do this, because of known gaps between these measures, it has been shown that some progress is, in principle, possible through the use of pseudo-complements. For a specific class of these, called slice functions, extremely close relationships exist between monotone and combinational complexity. These extend so far that we can identify specific monotone Boolean functions with the property that proofs of superpolynomial lower bounds on their monotone complexity would yield a proof that $P \neq NP$.

6. Further reading

Further work on slice functions, that has not been discussed in detail above, may be found in Wegener (1985, 1986) where, among other results, a set-theoretic model of computation based on slice functions is considered; and Dunne (1986, 1989). Dunne (1989) examines the problem of recomputing $f \in M_n$ given its $n + 1$ slice functions using only monotone operations, cf. Theorem 4.1(i) which uses negation and shows that, in this case, $O(n)$ gates suffice. Various positive (cases where savings can be made) and negative (cases where savings are not possible) results are obtained, the latter results giving exponential gaps.

An alternative approach to relating monotone and non-monotone complexity has been taken in Ugolnikov (1987). A generalisation of this work is described in Dunne (1988) pp. 263-268.

7. Appendix — Another Application of Theorem 3.5

Let $f \in B_N$ with formal arguments $\mathbf{X}_n^U$, i.e. f is a predicate on undirected n-vertex graphs. The function $CONN \in M_N$ is defined so that $CONN(\alpha) = 1$ if and only if the undirected graph $G(\alpha)$ is connected.

Lemma 7.1: If g in Theorem 3.5 is chosen to be $CONN$ then $h = CONN \wedge T_n^N$

Proof: From the statement of Theorem 3.5 we have that

$$h(\mathbf{X_n^U}) = \bigvee_{1 \le i < j \le n} (x_{i,j} \wedge CONN^{|x_{i,j} := 0}(\mathbf{X_n^U} - \{x_{i,j}\}))$$

We show that $h(\alpha) = 1$ if and only if $G(\alpha)$ is connected and contains at least n edges. Suppose $h(\alpha) = 1$. Then there is some $\{i, j\}$ such that

$$(x_{i,j} \wedge CONN^{|x_{i,j} := 0}(\mathbf{X_n^U} - \{x_{i,j}\}))(\alpha) = 1 \qquad \text{(i)}$$

This can only be true if the graph $G(\alpha)$ contains the edge $\{i, j\}$ and remains connected if this edge is deleted. Since any connected n-vertex graph has at least $n - 1$ edges it follows that $G(\alpha)$ has at least n edges and is connected. On the other hand if $G(\alpha)$ is connected and has at least n edges then there must be some edge $\{i, j\}$ in G whose removal does not disconnect G. It follows that there is a term of the form (i) in the definition of h.

The lemma is immediate since the function in the statement is true if and only if the corresponding graph is connected and has at least n edges. $\square$

Corollary 7.1: For $f \in B_N$, the function $F(f, CONN)(\mathbf{X_n^U})$ of Theorem 3.5 is:

$$F(f, CONN)(\mathbf{X_n^U}) = CONN(\mathbf{X_n^U}) \wedge (f \vee T_n^N)(\mathbf{X_n^U}) \qquad \square$$

Note that this function is *monotone*. For, suppose that $F(f, CONN)(\alpha) = 1$ and $x_{i,j} = 0$ in α. Then $F(f, CONN)(\beta)$, where β is obtained from α by increasing $x_{i,j}$, is also 1 since $CONN(\beta) = 1$ ($CONN$ is monotone) and at least n variables take the value 1 under β, hence $T_n^N(\beta) = 1$. We shall denote the function $F(f, CONN)$ by $Tree(f)$.

For any $f \in B_N$, $Tree(f)$ behaves as follows: $Tree(f)(\alpha) = 0$ if $G(\alpha)$ is not connected; $Tree(f)(\alpha) = 1$ if $G(\alpha)$ is connected and contains at least n edges; and $Tree(f)(\alpha) = f(\alpha)$ if $G(\alpha)$ is a *tree*. Thus $Tree(f)$ is interesting for those Boolean function corresponding to decision problems on n-node trees and gives a *monotone* 'encoding' of such properties. The 'standard' monotone encoding of trees with some property, Π say, (in which the prime implicants correspond to an enumeration of n-node trees with property Π) would only capture the decision problem 'does a given graph contain a *spanning tree* with property Π'. It is often the case that detecting if a tree has a given property is easier than detecting if a graph has a spanning tree with the same property, e.g. 'has at least k leaves for a given k', cf. Garey and Johnson (1979), p. 206.

For the remainder of this section we concentrate on decision problems for n-vertex trees. We first dismiss one possible objection: that adjacency matrices, i.e. $\mathbf{X_n^U}$, are not an efficient encoding of trees since $O(n^2)$ bits are used to encode $n-1$ edges.

Definition 7.1: Let T_n denote the set of all n-vertex labelled trees and $\sigma_n : T_n \rightarrow (\{0, 1, \# \}^*)$ be an encoding function that represents $T \in T_n$ by binary strings separated by $\#$ symbols. We say that σ is a *reasonable* encoding if the sequence of functions $\sigma = \langle \sigma_n \rangle$ is polynomial time computable (hence $|\sigma_n(T)|$ is bounded by a polynomial in n). We say that two encodings σ and τ are *compatible* if there are polynomial (in n) time algorithms to convert from $\sigma(T)$ to $\tau(T)$ and vice versa. •

Fact 7.1: If σ and τ are both reasonable encodings then σ and τ are compatible. □

Corollary 7.2: Let P be any decision problem on n-vertex trees, $\langle f_N \rangle$ be the corresponding family of Boolean functions in which trees are encoded by an adjacency matrix, and $\langle \sigma - f_M \rangle$ be the corresponding family in which trees are encoded by a reasonable encoding σ using a total of M binary inputs. Then there exist constants k and l such that:

$$\mathbf{C}(\sigma - f_M) \leq \mathbf{C}(Tree(f_N)) + O(n^k)$$

$$\mathbf{C}(Tree(f_N)) \leq \mathbf{C}(\sigma - f_M) + O(n^l)$$

Proof: Since $\mathbf{X_n^U}$ is clearly a reasonable encoding one can construct polynomial size combinational networks to convert between this any other reasonable encoding. □

We can now prove the main result of this section.

Theorem 7.1: Let P be any decision problem on trees. The combinational and monotone complexities of Boolean functions corresponding to P differ by at most an additive polynomial term.

Proof: Let $\langle f_N \rangle$ be the family of Boolean functions corresponding to P and $Tree(f)$ be the monotone encoding. From Theorem 3.5 and Lemma 7.1 it follows that $CONN^{|x_{i,j}=1} \wedge T_{n-1}^{N-1}$ is a pseudo-complement for $x_{i,j}$ when computing $Tree(f)$. Hence

$$\mathbf{C^m}(Tree(f)) = O(\mathbf{C}(Tree(f) + n^5)$$

by using $O(n^2)$ $CONN$ networks of size $O(n^3)$. The theorem now follows from Corollary 7.2. □

An interesting application of Theorem 7.1 concerns the problem below.

Definition 7.3: Given an n-vertex graph, $G(V, E)$ and natural number k ($1 \leq k \leq n$) as input, the *bandwidth problem* if to determine if there exists a bijective function $\beta : V \longleftrightarrow \{1, 2, \ldots, n\}$ such that $|\beta(v) - \beta(w)| \leq k$ for all $\{v, w\} \in E$. •

The bandwidth problem is NP-complete even if G is required to be a tree, Garey and Johnson (1979), p. 200.

If $P \neq NP$ then there must be some specific value of k for which the bandwidth problem does not have a polynomial time algorithm. Let $Band_k(\mathbf{X}_n^U)$ be the Boolean function corresponding to deciding if a tree has bandwidth at most k. From the preceding arguments, the following theorem is immediate.

Theorem 7.2: $P \neq NP$ if $Tree(Band_k)$ has superpolynomial monotone complexity for some k. $\square$.

References

Alon, N; Boppana, R: (1986) The monotone circuit complexity of Boolean functions; Combinatorica, 7, 1-22

Andreev, A.E: (1985) A method of proving lower bounds on the complexity of monotone Boolean functions; Doklady Akademii Nauk SSSR, 282, 1033-1037 (In Russian) (Transl: Sov. Math.-Doklady, 31, 530-534)

Andreev, A.E: (1987) A method of proving effective lower bounds on monotone complexity; Algebra and Logic, T.26, 1, 3-26 (In Russian)

Andreev, A.E: (1988) On the synthesis of schemes of functional elements from the monotone basis; Matem. Vopr. Kibernet. (Mathematical Results in Cybernetics), 1, 115-139 (In Russian)

Berkowitz, S: (1982) On some relationships between monotone and non-monotone circuit complexity; Technical Report, Univ. of Toronto

Blum, N: (1984a) A Boolean function requiring $3n$ network size; Theoretical Computer Science; 28, 337-345

Blum, N: (1984b) An $\Omega(n^{4/3})$ lower bound on the monotone network complexity of n-th degree convolution; Theoretical Computer Science, 36, 59-70

Bollobás, B: (1978) Extremal Graph Theory; Academic Press

Borodin, A: (1977) On relating time and space to size and depth; SIAM Jnl. on Computing, 6, 733-744

Cook, S.A: (1971) The complexity of theorem proving procedures; Proc. 3rd ACM Symposium on Theory of Computing, 151-158

Dedekind, R: (1897) Uber Zerlegungen von Zahlen durch ihre grossten gemeinsamen Teiler; Reprinted in: Ges. Math. Werke II, Chelsea, N.Y (1969), 103-108

Dunne, P.E: (1984) Techniques for the analysis of monotone Boolean networks; Ph.D Dissertation; Theory of Computation Report No.69, Dept. of Comp. Sci., Univ. of Warwick,

Dunne, P.E: (1985) A $2.5n$ lower bound on the monotone network complexity of T_3^n; Acta Informatica, 22, 229-240

Dunne, P.E: (1986) The complexity of central slice functions; Theoretical Computer Science, 44, 247-257

Dunne, P.E: (1988) The Complexity of Boolean Networks; Academic Press

Dunne, P.E: (1989) On monotone simulations of non-monotone networks; Theoretical Computer Science, 66, 15-25

Erdös, P; Simonovits, M: (1973) On a valence problem in extremal graph theory; Discrete Math., 5, 323-334

Fischer, M; Pippenger, N.J: (1979) Relations among complexity measures; Jnl. of the ACM, 26, 361-381

Garey, M; Johnson, D: (1979) Computers and intractability - a guide to the theory of NP-completeness; Freeman

Gaskov, S.B: (1978) The depth of Boolean functions; Problemy Kibern., 34, 265-268 (In Russian) (English description: Dunne (1988), pp. 58-63) (NB: this paper assumes considerable familiarity with Lupanov (1973))

Gilbert, E.N: (1954) Lattice theoretic properties of frontal switching functions; Jnl. Math. and Phys., 33, 57-97

Hopcroft, J.E; Karp, R.M: (1973) An $n^{5/2}$ algorithm for maximum matching in bipartite graphs; SIAM Jnl. on Computing, 2, 225-231

Jerrum, M; Snir, M: (1982) Some exact complexity results for straight-line computations over semirings; Jnl. of the ACM, 29, 874-897

Karchmer, M; Wigderson, A: (1987) Monotone circuits for connectivity require super-logarithmic depth; Internal Report, Hebrew Univ., Jerusalem

Korshunov, A.D: (1981) On the number of monotone Boolean functions; Problemy Kibernet., 38, 5-108 (In Russian) (Brief English summaries in: Dunne (1988), pp. 124, 146-147; Wegener (1987), p. 99)

Lupanov, O.B: (1958) On a method of circuit synthesis; Izvestia VUZ (Radiofizika), 1, 120-140 (In Russian) (English descriptions in: Dunne (1988), pp. 45-50; Savage (1976), pp. 116-119; Wegener (1987), pp. 91-92)

Lupanov, O.B: (1973) Complexity of the universal parallel-series network of depth 3; Trudy Matem. Inst. Steklov, 133, 127-131 (In Russian) (English description in: Dunne(1988), pp. 58-61, 360-363)

Mehlhorn, K; Galil, Z: (1976) Monotone switching networks and Boolean matrix product, Computing, 16, 99-111

Paterson, M.S: (1975) Complexity of monotone networks for Boolean matrix product; Theoretical Computer Science, 1, 13-20

Post, E.L: (1941) Two-valued iterative systems of mathematical logic; Annals of Math. Studies, 5, Princeton Univ. Press

Raz, R; Wigderson, A: (1990) Monotone circuits for matching require linear depth; Internal Report, Hebrew Univ., Jerusalem

Razborov, A.A: (1985a) Lower bounds on the monotone complexity of some Boolean functions; Doklady Akademii Nauk SSSR, 281, 798-801; (In Russian) (Transl: Sov. Math. Doklady, 31, 354-357)

Razborov, A.A: (1985b) A lower bound on the monotone complexity of the logical permanent; Mat. Zametki, 37, 887-901; (In Russian) (Transl: Mathem. Notes of the Acad. of Sci. of the USSR, 37, 485-493)

Razborov, A.A: (1988) An application of matrix methods to the theory of lower bounds on the complexity of computation; Preprint, Steklov Institute, Moscow Univ. (In Russian)

Red'kin, N.P: (1979) On the realisation of monotone Boolean functions by contact circuits; Problemy Kibernet., 35, 87-110 (In Russian) (English description in: Dunne (1988), pp. 140-146)

Riordan, J; Shannon, C.E: (1942) The number of two-terminal series-parallel networks; Jnl. Math. and Phys. 21, 83-93

Savage, J.E: (1972) Computational work and time on finite machines; Jnl. of the ACM, 19, 660-674

Savage, J.E: (1976) The Complexity of Computing; Wiley

Schnorr, C.P: (1976a) The network complexity and Turing machine complexity of finite functions; Acta Informatica, 7, 95-107

Schnorr, C.P: (1976b) A lower bound on the number of additions in monotone computations of monotone rational polynomials; Theoretical Computer Science, 2, 305-317

Shannon, C.E: (1949) The synthesis of two-terminal switching circuits; Bell System Tech. Jnl., 28, 59-98

Tardos, E: (1987) The gap between monotone and non-monotone circuit complexity is exponential; Combinatorica, 7, 141-142

Tiekenheinrich, J: (1984) A $4n$ lower bound on the monotone Boolean network complexity of a one output Boolean function; Inf. Proc. Letters, 18, 201-202

Ugolnikov, A.B: (1987) On the complexity of realising Boolean functions by schemes over the basis of majority and implication; Vestn. Mosc. Un-ta. Ser. 1, Matematika Mechanika, 4, 76-78 (In Russian) (English description in: Dunne (1988), pp. 263-267)

Valiant, L.G: (1986) Negation is powerless for Boolean slice functions; SIAM Jnl. on Computing, 15, 531-535

Van Voorhis, C.C: (1972) An improved lower bound for sorting networks; IEEE Trans. Computers, C-21, 612-613

Wegener, I: (1985) On the complexity of slice functions; Theoretical Computer Science, 38, 55-68

Wegener, I: (1986) More on the complexity of slice functions; Theoretical Computer Science, 43, 201-211

Wegener, I: (1987) The Complexity of Boolean Functions, Wiley-Teubner

Weiss, J: (1983) An $\Omega(n^{3/2})$ lower bound on the complexity of Boolean convolution; Information and Control; 59, 84-88

On Read-Once Boolean functions

Ilan Newman *

Abstract

We survey some recent results on read-once Boolean functions. Among them are a characterization theorem, a generalization and a discussion on the randomized Boolean decision tree complexity for read-once functions. A previously unpublished result of Lovás and Newman is also presented.

1. Introduction

A *Boolean formula* is a rooted binary tree whose internal nodes are labeled by the Boolean operators $\vee$ or $\wedge$ and in which each leaf is labeled by a Boolean variable or its negation. A Boolean formula computes a Boolean function in a natural way.

A Boolean formula is *read-once* if every variable appears exactly once. A function is *read-once* if it has a read-once formula.

Read-once functions have been studied by many authors, since they have the lowest possible formula size (for functions that depend on all their variables). In addition, every NC^1 function on n variables is a projection of a read-once function with a polynomial (in n) number of variables.

We present here some recent results in the area. All but one of those results have been published, hence, full proofs will be generally omitted and will be given just for the unpublished result (Theorem 3.4). The results we will discuss cover a characterization theorem, some generalizations and results on the randomized decision tree complexity of read-once functions. There is a recent result on learning of read-once functions, [AHK89], which will not be described.

---

*KTH, NADA S-100 44 Stockholm, Sweden. This work was done while the author was at the Hebrew University, Jerusalem.

2.   Definitions and Notations

If $g : \{0,1\}^n \mapsto \{0,1\}$ has a formula in which no negated variable appears, we say that g is *monotone*. The size of a Boolean formula is the number of its leaves. Its depth is the depth of the longest path from the root to a leaf. The formula size of a Boolean function g, denoted $L(g)$, is the minimum formula size over all the formulae that compute g. Similarly, the depth of a Boolean function, denoted $depth(g)$, is the minimum depth of a formula over all formulae that compute the function. For a monotone Boolean function, monotone size and depth, denoted $L_m(g)$ and $depth_m(g)$, are defined similarly by taking the minimization over monotone formulae. A theorem of Spira, [Spi71], shows that for any Boolean function g, $depth(g) = O(\log L(g))$ and $depth_m(g) = O(\log L_m(g))$.

Let $g : \{0,1\}^n \mapsto \{0,1\}$ be a monotone Boolean function on n variables. We identify the variable set with the set $\{1, ..., n\} = [n]$. A *minterm* of g is a minimal set $S \subseteq [n]$ such that if we assign '1' to the variables in S it forces the function to be '1'. We denote by $Min(g)$ the set of all minterms of g. Similarly, a *maxterm* of g is a minimal set $T \subseteq [n]$, such that if we assign '0' to the variables in T it forces the function to be '0'. We denote by $Max(g)$ the set of all maxterms of g.

We say that g has the *t-intersection* property if for every minterm P of g and every maxterm Q of g, $|P \cap Q| \le t$.

We note here several facts about minterms and maxterms of monotone functions.

1. For any monotone Boolean function g,

$$\forall S \in Min(g), \ \forall T \in Max(g), \ S \cap T \ne \phi$$

2. For any antichain $\mathcal{C} \subseteq 2^{[n]}$ there is a monotone function $g : \{0,1\}^n \mapsto \{0,1\}$ such that $Min(g) = \mathcal{C}$. Moreover, in that case

$$Max(g) = \{T| \ \forall S \in \mathcal{C}, |S \cap T| \ge 1, \ and \ T \ is \ minimal\}$$

A dual statement for $Max(g) = \mathcal{C}$ is also true.

3. Characterization of Read-Once Functions and generalizations

3.1. Characterization

A characterization theorem for read-once functions has been proved first by Gurvich [Gur77, Gur82] and independently by other authors, [Mun89, BP90, KLNSW88]. It can be deduced also from [Sey76, Sey77].

Theorem 3.1 *[Gur77, Gur82] Let $g : \{0,1\}^n \mapsto \{0,1\}$ be a monotone Boolean function that depends on all its variables, then g is read-once if and only if g has the 1-intersection property.*

We here sketch the proof presented by Karchmer et al. in [KLNSW88]. The proof that a read-once function has the 1−intersection property is easy, and can be shown by top down induction on the structure of the read-once formula for the function. The opposite direction is more involved.

For a monotone Boolean function $g : \{0,1\}^n \mapsto \{0,1\}$, define its *minterm (maxterm) graph* to be $G = (V,E)$ where $V = [n]$ and two variables i,j are connected by an edge if there is a minterm (maxterm) of g that contains both.

The proof that a function g that has the 1-intersection property is read-once goes by induction on n, the number of variables of g. It can be shown that the set of maximal cliques of the minterm graph G of g is $Min(g)$ and the set of maximal independent sets of G is $Max(g)$. It follows that the maxterm graph is the complement of the minterm graph. Further more, using this and the 1-intersection property, it follows that either this graph is disconnected or its complement (the maxterm graph) is disconnected. Say, the minterm graph is disconnected. Define a Boolean function, g_i, for the i-th connected component of G, to be the function whose minterms are the maximal cliques of that component. It can be seen that since g has the 1-intersection property so does g_i for every i. So, by induction hypothesis, the g_i's are read-once, and since $g = \vee_i g_i$, g is read-once too. If the case is that the maxterm graph is disconnected, a similar argument shows that $g = \wedge_i h_i$ where h_i's are the read-once functions obtained by a similar construction on the maxterm graph.

The characterization theorem can be generalized in a straightforward manner to the non-monotone case, using the proper definitions of minterms and maxterms. It appears in [KLNSW88] but will not be stated here.

3.2. Generalization to Read-Once On a Subset of the Variables

Using [KLNSW88], L. Hellerstein [He90] generalized the characterization of read-once functions to functions that are read-once on a subset of their variables.

Definition 3.2 *A monotone Boolean function* $g : \{0,1\}^n \mapsto \{0,1\}$ *is read-once on a subset Z of its variables if it has a formula in which every variable in Z appears exactly once.*

For a monotone Boolean function g denote its minterm graph by G_g and its maxterm graph by H_g. For any graph G and $V' \subseteq V(G)$, $G(V')$ is the induced subgraph on V'.

Theorem 3.3 *[He90] A monotone Boolean function* $g : \{0,1\}^n \mapsto \{0,1\}$ *is read-once on a subset $Z \subseteq [n]$ if and only if $\forall Z' \subseteq Z$, $|Z'| > 1$ either $G_g(Z')$ or $H_g(Z')$ is disconnected.*

A read-once function is clearly read-once on any subset of its variables. The proof of Theorem 3.1 given in [KLNSW88] shows that Theorem 3.3 holds for the case of $Z = [n]$ (read-once functions). Indeed, if $G_g(Z')$ or $H_g(Z')$ is disconnected for every $Z' \subseteq Z$, $(|Z| > 1)$ one can inductively express g as $g = \vee_i g_i$ or $g = \wedge_i g_i$ where each g_i is read-once as in the proof sketch of Theorem 3.1 given before. (The other direction follows directly by induction on the structure of the formula).

3.3. Functions with the t-intersection property.

The characterization theorem for read-once functions asserted that if a function has the 1-intersection property then its formula size is 'small'. We want to relate the t-intersection property to the formula size of the corresponding Boolean function for general t. The following result is a previously unpublished theorem due to L. Lovász and I. Newman.

Theorem 3.4 *Let* $g : \{0,1\}^n \mapsto \{0,1\}$ *be a monotone Boolean function that has the t-intersection property, then g has a monotone formula of size* $O(n^{t^2 \log n})$.

A full proof of Theorem 3.4 is presented here. The proof relies upon relating the formula size of g to the communication complexity of an appropriate communication game, using the Karchmer-Wigderson Theorem [KW88], and applying a previous result of Lovász and Saks [LS88]. Lovász and Saks's theorem is first stated without proof as Theorem 3.5 below. Some preliminary definitions are required.

Let f be a Boolean function $f : X \times Y \mapsto \{0,1\}$. The *communication problem* of computing f is the scene where we have two parties, A and B. A gets $x \in X$, B gets $y \in Y$ and they have to communicate according to a predetermined protocol in order to compute $f(x,y)$. The *communication complexity* of such a protocol is the number of bits transferred between the parties (in the worst case of x, y).

For f as above we associate a 0-1 matrix $M(f)$ whose rows are indexed by $x \in X$, columns are indexed by $y \in Y$ and $M(f)_{x,y} = f(x,y)$. A *0-rectangle* of $M(f)$ is a submatrix defined by some $X' \subseteq X$ and $Y' \subseteq Y$, in which all entries are 0. A *0-cover* of $M(f)$ is a collection of 0-rectangles such that any 0-entry of $M(f)$ is in one of the rectangles.

For a matrix M (not necessarily 0-1), define the *triangle-rank* of M, denoted by $trank(M)$, to be the size of biggest nonsingular lower triangular submatrix of M. Clearly $rank(M) \geq trank(M)$.

Theorem 3.5 *[LS88] Let* $f : X \times Y \mapsto \{0,1\}$. *Suppose* $M(f)$ *has a 0-cover consisting of* c *rectangles, and that* $trank(M(f)) \leq r$. *Then, there is a protocol for computing f, of complexity* $O((\log c)(\log r))$.

Remark: Implicitly in the proof of [LS88], if the protocol, when run on a pair (x,y) ends with a '0' answer, then both players know a 0-rectangle from the given 0-cover, in which the (x,y) entry of $M(f)$ is contained.

Proof of Theorem 3.4 : We use the Karchmer-Wigderson Theorem [KW88] to prove that there is a (monotone) formula of depth $O((t \log n)^2)$ that computes g. The Karchmer-Wigderson theorem relates a communication game of two players to the depth of the formula. Specifically, we have two players Min and Max. Min gets a minterm P of g, Max gets a maxterm Q of g. They have to communicate according to a predetermined protocol in order to find an element $i \in P \cap Q$. The communication complexity of a given protocol is the number of bits transferred between the two players in the worst case choice of P, Q. The communication complexity of the game is the communication complexity of the best protocol.

Karchmer and Wigderson proved that this complexity measure is always equal to the monotone depth of the function.

We show that in our particular case there is a protocol for the appropriate communication game of complexity $O((t \log n)^2)$.

Let $g : \{0,1\}^n \mapsto \{0,1\}$ be a monotone Boolean function with the t-intersection property. Define
$f_t : Min(g) \times Max(g) \mapsto \{0, ..., t-1\}$ by $f_t(P, Q) = t - |P \cap Q|$,
$h_t : Min(g) \times Max(g) \mapsto \{0,1\}$ by $h_t(P, Q) = 0$ if $f_t(P, Q) = 0$ and 1 otherwise.

h_t is a Boolean function and $M(h_t)$ has a 0-cover that consists of at most $\binom{n}{t}$ rectangles. To see that, observe that for any t-element set T, the set $R_T = \{P | P \in Min(g), T \subseteq P\} \times \{Q | Q \in Max(g), T \subseteq Q\}$ is a 0-rectangle of $M(h_t)$, and that any 0-entry of $M(h_t)$ is in such a rectangle.

Moreover, we have $trank(M(h_t)) = trank(M(f_t)) \leq rank(M(f_t)) \leq n + 1$. The last equality being true since $M(f_t) = tJ - U \times W^T$, where J is the all-1 matrix, U is the incidence matrix of minterms (i.e., of dimension $|Min(g)| \times n$ for which the (P, i) entry is '1' if $i \in P$ and '0' otherwise), and W is the incidence matrix of maxterms (defined similarly).

Thus, by Theorem 3.5, we have a protocol for h_t of complexity $O(\log n \log \binom{n}{t})$ $= O(t(\log n)^2)$. By the remark, if the players get a '0' answer, they also know a t-element set that is contained in both P and Q (a 0-rectangle). They can decide on, say, the minimal element in that t-set as the answer to the original game.

In the other case, if the answer to $h_t(P, Q)$ is '1', both players know of a submatrix $X' \times Y'$, such that $X' \subseteq Min(g), Y' \subseteq Max(g), |P' \cap Q'| \leq t - 1$ for all $P' \in X'$, $Q' \in Y'$, and $P \in X'$, $Q \in Y'$. So, they can define f_{t-1}, h_{t-1} accordingly, and repeat the process. This can be repeated at most t times, after which the players find an element in the intersection.

The whole protocol is thus, of complexity $O((t \log n)^2)$. $\square$

4.   Read-Once Functions and the Randomized Boolean Decision Tree Model

A *Boolean Decision Tree* T is a rooted binary tree, each internal node is labeled by a Boolean variable and the two out going edges are labeled by '0' or '1'. Each leaf is also labeled '0' or '1'. A decision tree defines a Boolean function in a natural way: Every input (a 0-1 assignment of the variables) defines a unique path from the root to a leaf by starting at the root and branching according to the values of the input variables (on each branch we say that a *query* to the corresponding variable is being made). The value of the function on that input is the label of the leaf in the end of this path. The complexity of the tree on an input is the length of the corresponding path (i.e the number of queries along this path). The complexity of a given tree is its depth, that is, the number of variables being queried for the worst case input. *The decision tree complexity* of a Boolean function $g : \{0,1\}^n \mapsto \{0,1\}$, denoted $D(g)$, is the minimum complexity over all decision trees that define g.

The decision tree model was studied by many authors due to its simplicity and its relation to some other natural properties of a Boolean function. See, [Yao77, RV78, KSS84, SW86, Yao87, Kin88, Haj90, Nis89, HNW90, HW91].

A randomized version has also been defined, see [Yao77, SW86] for exact definitions and discussion. Intuitively, a randomized decision tree for g is a probabilistic procedure of querying variables' values in order to evaluate g. Formally, it is a probability distribution over deterministic decision trees for g. The complexity of a randomized decision tree for an input is the average length of the paths it defines in the individual deterministic decision trees. The complexity of the randomized decision tree on the function is its randomized complexity for the worst case input. Finally, for a Boolean function $g : \{0,1\}^n \mapsto \{0,1\}$, its *randomized decision tree complexity*, denoted here by $RD(g)$, is the minimum complexity over all randomized decision trees for g.

A well known fact in this model is that for any Boolean function $g : \{0,1\}^n \mapsto \{0,1\}$, $RD(g) = \Omega(D(g)^{0.5})$. An interesting open problem is whether this bound is tight. This question was partially answered for monotone graph properties in [Yao87, Kin88, Haj90].

It turns out that read-once functions are easier to deal with in proving lower bounds on the randomized decision tree complexity. We want to state here some results in that direction.

It is easy to see that every read-once function $g : \{0,1\}^n \mapsto \{0,1\}$ has $D(g) = n$ (which is the highest possible). Saks and Wigderson [SW86] proved a lower bound on the randomized decision tree complexity for the 'canonical' NC^1 function, defined hereafter.

Definition 4.1 *For $n = 2^m$ let $g : \{0,1\}^n \mapsto \{0,1\}$ be a read-once function defined by the complete binary tree formula of depth m in which the odd level gates are $\wedge$ and the even level gates are $\vee$.*

Theorem 4.2 *[SW86] $RD(g) = \Omega(n^\alpha)$ where $\alpha = \log \frac{1+\sqrt{33}}{4}$.*

Saks and Wigderson conjectured that among all Boolean functions, the gap between $RD(g)$ and $D(g)$ is the biggest possible for the above function. This conjecture is still open. Heiman and Wigderson proved that the mentioned gap between the deterministic and the randomized decision tree complexity for read-once functions is smaller then quadratic.

Theorem 4.3 *[HW91] There is a constant ϵ such that for any read-once function $g : \{0,1\}^n \mapsto \{0,1\}$ $RD(g) = \Omega(n^{0.5+\epsilon})$.*

Heiman et al. [HNW90] used the same methods to prove a lower bound for the randomized decision tree complexity for bounded depth threshold read-once functions. The statement of that theorem will not be given here.

We do not give the proofs of those results here, we just note that all the results above were proved by using an inductive argument developed by Saks and Wigderson [SW86] and a general lemma of Yao [Yao77] that relates randomized complexity to 'average case' complexity. For details see [SW86].

Acknowledgments
I would like to thank Avi Wigderson for many helpful discussions. I am also indebted to the anonymous referee that suggested many style improvements and informed me on a reference I was not aware of.

References

[AHK89]   Angluin D., Hellerstein L., Karpinski M., *Learning Read-Once Formulas with Queries. UC Berkeley Technical Report UCB CSD 89-528, to appear in JACM.*

[BP90]   Beynon M., Paterson M., *1990, Personal communication.*

[Gur77]   Gurvich V. A., *On repetition-Free Boolean Functions.* sl Uspekhi Matematicheskikh Nauk, 1977 V. 32, (1) 183-184. (in Russian).

[Gur82]   Gurvich V. A., *On the Normal Form of positional Games. Soviet Math. Dokl. Vol 25, No. 3, (1982) 572-574.*

[Haj90]   Hajnal P., *On the power of randomness in the decision tree model. Proc. 5th Annual Symp. on Structures in Complexity Theory, (1990) 66-77.*

[He90]   Hellerstein L., *Functions that are Read-Once on a Subset of their Inputs*. to appear in Discrete Applied Math.

[HNW90]   Heiman R., Newman I., Wigderson A., *On read-once threshold formulae, and their randomized decision tree complexity*. Proc. 5th Annual conference on Structures in Complexity Theory, (1990) 78-87.

[HW91]   Heiman R., Wigderson A., *Randomized vs. deterministic decision tree complexity for read-once Boolean functions*. Proc. of Structures 1991, pp 172-179.

[Kin88]   King V., *Lower bounds on the complexity of graph properties*. Proc. 20th ACM Symp. on Theory of Computing, 1988, pp. 468-476.

[KLNSW88]   Karchmer M., Linial N., Newman I., Saks M., Wigderson A., *Combinatorial Characterization of Read-Once Formulae*. to appear in Discrete Math.

[KSS84]   Kahn J., Saks M., Sturtevant D., *A topological approach to evasiveness*. Combinatorica 4 (1984), 297-306.

[KW88]   Karchmer M., Wigderson A., *Monotone circuits for connectivity require super- logarithmic depth*. Proc. 20th Annual ACM Symp. on Theory of Computing, (1988) 539-550.

[LS88]   Lovász L., Saks M., *Lattices, Mobius function and Communication Complexity*. Proc. 29th Annual Symp. on Foundation of Computer Science, (1988) 81-90.

[Mun89]   Mundici D., *Functions computed by monotone Boolean formulas with no repeated variables*. J. Theoretical Computer Science 66 (1989) 113-114.

[Nis89]   Nisan N., *CREW PRAMs and decision trees*. Proc. 21st ACM Symp. on Theory of Computing, 1989, pp. 327-335.

[RV78]   Rivest R., Viullemin S., *On recognizing graph properties from adjacency matrices*. Theoretical Computer Science 3, 1978, pp. 371-384.

[Sey76]   Seymour P.D., *The forbidden minors of binary clutters*. J. London Mathematical Soc. 12(1976) 356-360.

[Sey77]   Seymour P.D., *Note, A note on the production of matroids minors*. J. Combinatorial Theory B, 22(1977) 289-295.

[Spi71]   Spira P.M., *On Time-Hardware Complexity Tradeoffs for Boolean Functions*. Proc. 4th Hawaii Symp. on System Sciences, 1971, 525-527.

[SW86]   Saks M., Wigderson A., *Probabilistic Boolean decision trees and the complexity of evaluating game trees*. Proc. 27th IEEE Symp. on Foundations of Computer Science, 1986, pp. 29-38.

[Yao77]   Yao A. C.,*Probabilistic computation, towards a unified measure of complexity*. Proc. 18th IEEE Symp. on Foundation of Computer Science, 1977, 222-227.

[Yao87]   Yao A. C., *Lower bounds to randomized algorithm for graph properties*. Proc. 28th IEEE Symp. on Foundations of Computer Science, 1987, pp. 393-400.

Boolean Function Complexity: a Lattice-Theoretic Perspective

W. Meurig Beynon [*]

Abstract

Topical but classical results concerning the incidence relationship between prime clauses and implicants of a monotone Boolean function are derived by applying a general theory of computational equivalence and replaceability to distributive lattices. A non-standard combinatorial model for the free distributive lattice FDL(n) is described, and a correspondence between monotone Boolean functions and partitions of a standard Cayley diagram for the symmetric group is derived.

Preliminary research on classifying and characterising the simple paths and circuits that are the blocks of this partition is summarised. It is shown in particular that each path and circuit corresponds to a characteristic configuration of implicants and clauses. The motivation for the research and expected future directions are briefly outlined.

1. Introduction

Models of Boolean formulae expressed in terms of the incidence relationship between the prime implicants and clauses of a function were first discovered several years ago [14], but they have recently been independently rediscovered by several authors, and have attracted renewed interest. They have been used in proving lower bounds by Karchmer and Wigderson [17] and subsequently by Razborov [21]. More general investigations aimed at relating the complexity of functions to the model have also been carried out by Newman [20].

This paper demonstrates the close connection between these classical models for monotone Boolean formulae and circuits and a general theory of computational equivalence as it applies to FDL(n): the (finite) distributive lattice

---

[*]Department of Computer Science, University of Warwick, Coventry, CV4 7AL, England. The author was partially supported by the ESPRIT II BRA Programme of the EC under contract # 3075 (ALCOM).

freely generated by n elements. It also describes how the incidence relationships between prime implicants and clauses associated with monotone
Boolean functions can be viewed as built up from a characteristic class of
incidence patterns between relatively small subsets of implicants and clauses.
Each incidence pattern is associated with a sequence of permutations resem-
bling the *circular sequences* previously studied in computational geometry
[12]. These sequences of permutations are derived via a combinatorial model
for the free distributive lattice FDL(n) in which a monotone Boolean function is represented by a map from the symmetric group Σ_n onto the set
$\{1, 2, \ldots, n\}$ — a *combinatorially piecewise-linear* (cpl) map. Each cpl map
determines a set of singular chains and cycles; these are simple paths and
circuits partitioning the Cayley diagram Γ_n of Σ_n as generated by the set of
transpositions of adjacent symbols.

The work described in this paper contributes to a broad programme of research into richer algebraic foundations for Boolean function complexity. The
theory of computational equivalence shows that computation with monotone
Boolean functions (i.e., in the free distributive lattice FDL(n)) has a natural
generalisation to finite distributive lattices. This is helpful in understanding
the relationship between monotone and non-monotone computational models
(as represented especially in the work of Dunne [11] on pseudo-complements).
It also establishes links with previous work of the author that motivated the
introduction of cpl maps by relating the theory of distributive lattices to
piecewise-linear geometry [3]. The final section of the paper briefly explains
how this link with geometry is currently being developed.

The paper is in two main parts. Both parts of the paper are more fully developed elsewhere; refer to [4, 5] for more background and for proofs of theorems,
where omitted. The first part reviews results on computational equivalence
and replaceability and derives models for formulae and circuits. The second
introduces the combinatorial model of FDL(n) and explains how the paths
and cycles in Γ_n arise. The problem of classifying and characterising singular
chains and cycles is addressed from a group-theoretic and a lattice-theoretic
perspective. It is shown informally that the number of equivalence classes
of singular chains and cycles is small relative to the number of monotone
Boolean functions, and that each cycle corresponds to a characteristic configuration of implicants and clauses. The overall context for the research, and
some issues and directions for future work are also briefly described.

2. Boolean computation: a lattice-theoretic view

Boolean function theory has been developed in the two cultures of algebra and of complexity theory. This paper adopts an algebraic perspective, focusing primarily upon monotone Boolean functions and the associated theory of finite distributive lattices. The classical algebraic structure theory of finite distributive lattices and Boolean algebras does not provide a basis for understanding the complexity of Boolean functions. Possible ways of redeeming this situation are part of the motivation for the work described in this paper.

Some basic algebraic concepts, such as are fully described in [8], are required. A *distributive lattice* is an algebra with respect to two operations, meet ($\wedge$) and join ($\vee$), that are associative, idempotent and commutative and are mutually distributive. The classical representation theory of finite distributive lattices is very elementary: every finite distributive lattice is uniquely determined by its poset of join-irreducible (or dually, meet-irreducible) elements. Indeed, a finite distributive lattice is essentially isomorphic with the lattice of decreasing subsets of its poset of join-irreducibles, as partially ordered by set inclusion. A distributive lattice is *Boolean* if every element x has a complement x' such that $x \vee x'=1$ and $x \wedge x'=0$, where 1 and 0 respectively denote the top and bottom elements. The free distributive lattice on n generators is finite. Its poset of join-irreducibles is essentially the Boolean lattice with n atoms. Elementary as $FDL(n)$ appears, there is some evidence to indicate that it has a deceptively subtle structure. For example, it has proved hard to enumerate the elements of $FDL(n)$ for $n > 7$.

For the complexity theorist, an archetypal problem is the determination of bounds on the size of a circuit to compute a particular *monotone Boolean function*. In algebraic terms, a monotone Boolean function in the inputs $x_1, x_2, \ldots, x_n$ is an element of the finite distributive lattice $FDL(n)$ freely generated by elements $x_1, x_2, \ldots, x_n$. The disjunctive normal form of a monotone Boolean function f is the canonical representation of the element f as an irredundant join of join-irreducible elements in $FDL(n)$, and dually. In the algebraic perspective adopted in this paper, the computation of f is viewed as one of a more general class of computations concerned with computing an element of an arbitrary finite distributive lattice from a given subset of the lattice. In this generalisation, the role of prime implicants is played by arbitrary join-irreducibles. It will be shown that, in this more general computational framework, it is possible in principle to gain insight into general Boolean function complexity (see §2.3).

2.1. Computational equivalence and replaceability

Suppose that X is an algebra with signature Ω, generated as an Ω-algebra by the elements $x_1, x_2, \ldots, x_n$. If f, g and h are elements of X, then h is replaceable by g for the purpose of computing f, written $h \sqsupseteq_f g$, if for all Ω-expressions $w(z, z_1, z_2, \ldots, z_n)$:

$$w(h, x_1, x_2, \ldots, x_n) = f \implies w(g, x_1, x_2, \ldots, x_n) = f.$$

The relation $\sqsupseteq_f$ defines a pre-order (i.e., reflexive and transitive relation) on X respecting the operators in Ω. This pre-order induces a partial order upon a quotient of X, as follows:

Theorem 2.1 *Let $\square_f$ be the equivalence relation on X defined by*

$$\square_f \equiv \sqsupseteq_f \cap \sqsubseteq_f.$$

Then

(i) $\square_f$ respects the operations in Ω, so that the quotient algebra $X/\square_f$ is partially ordered by $\sqsupseteq_f$,

(ii) the element f is solitary under $\square_f$: that is, it is the unique element in its equivalence class,

(iii) $\square_f$ is characterised as the largest equivalence relation on X satisfying (i) and (ii).

The relation $\square_f$ is trivial in many cases — for instance, if X has cancellative properties, as in group theory. In such cases, the partial order $\sqsupseteq_f$ may nonetheless be non-trivial. A simple example of an algebra for which both $\square_f$ and $\sqsupseteq_f$ are non-trivial is the multiplicative group of residues modulo a composite integer. (For more illustrations and a proof of Theorem 2.1, consult [4, 7].)

It is of interest to note that the less powerful the algebraic structure imposed upon X by the operators in Ω, the more complex is the associated theory of computational equivalence and replaceability. This paper deals only with the theory as it applies to finite distributive lattices, but related theories have also been developed for general lattices [7, 10], and for semigroups [15, 22].

2.2. The case of distributive lattices

Replaceability is a concept that has been used effectively in arguments about monotone Boolean functions. The results on replaceability summarised here are generalisations to arbitrary finite distributive lattices of results that were first developed by Dunne (see [11] pp. 147-152) for monotone Boolean functions, i.e., for the case of FDL(n). For a full account of Theorems 2.2 and 2.3, including proofs and discussion of corollaries, see [4]. Computational equivalence relations on finite distributive lattices provide a more general and elegant algebraic framework within which to view replaceability relations.

The following theorem gives a characterisation of the quotient defined by computational equivalence in a finite distributive lattice:

Theorem 2.2 *Let f be an element of the finite distributive lattice D. There is an element $\lambda(f)$ in D characterised as the least element of D such that the interval $[\lambda(f), f]$ is a Boolean lattice. Dually, there is an element $\mu(f)$ in D characterised as the greatest element of D such that the interval $[f, \mu(f)]$ is a Boolean lattice. Then*

(i) $\lambda(f)$ is the unique maximal 0-replaceable element modulo f,

(ii) $D/\Box_f$ is isomorphic with $[\lambda(f), \mu(f)]$ via the canonical retract:

$$D \rightarrow [\lambda(f), \mu(f)] \text{ defined by mapping } x \text{ to } (x \vee \lambda(f)) \wedge \mu(f),$$

(iii) $D/\Box_f$ is an abstract simplicial complex (i.e., the lattice of faces of an n-dimensional geometric object composed of simplices) with respect to the partial ordering $\sqsupset_f$.

Theorem 2.2 supplies a structural perspective on computational equivalence, indicating how the computation of the element f in the lattice D can be viewed with reference to the algebraic structure of D. The elements of the form $\lambda(f)$ and $\mu(f)$ define closure lattices within D that have independent interest [4]. An alternative characterisation of computational equivalence and replaceability modulo f, more useful for computational purposes, is the following:

Theorem 2.3 *Let f be an element of the finite distributive lattice D. Let P_f denote the set of join-irreducibles in the unique irredundant representation of f as a join of join-irreducibles of D. Dually, let Q_f denote the set of meet-irreducibles in the unique irredundant representation of f as a meet of meet-irreducibles of D. If g and h are elements of D, then*

$$g \sqsupset_f h \text{ iff } (\forall p \in P_f \colon p \leq g \Rightarrow p \leq h) \text{ and } (\forall q \in Q_f \colon q \geq g \Rightarrow q \geq h).$$

2.3. Applications

In combination, the characterisations of computational equivalence and replaceability can be used to derive the explicit forms for replacement rules as specified by Dunne for the special case of $\mathrm{FDL}(n)$ [11]. These rules subsume many previous specific applications of replaceability arguments in defining bounds on circuit size. Following Berkowitz [2], it is known that in certain contexts non-monotone computations can be directly simulated in a monotone framework without significantly changing their complexity. The instances of such simulation discussed in [2] and [11] can be viewed as particular cases that arise when the computational equivalence lattice $\mathrm{D}/\square_f \cong [\lambda(f), \mu(f)]$, as determined by the two Boolean intervals $[\lambda(f), f]$ and $[f, \mu(f)]$, is itself a Boolean lattice. In the cases considered in [2] and [11], the simulation respects the complexity of the computation (to within a constant factor) because elements computationally equivalent to the complemented inputs can be computed efficiently.

It is easy to show that an algorithm to determine the complexity of arbitrary computational problems in finite distributive lattices would essentially be sufficient to determine the general Boolean complexity of a monotone function. It suffices to prove that if $f \in \mathrm{FDL}(n)$, then $\mathrm{FDL}(n + 1)/\square_f$ is Boolean. If p is a join-irreducible in a finite distributive lattice, the complement of the set of elements $\{x \mid x \leq p\}$ is the set of elements $\{x \mid x \geq \tilde{p}\}$, for some meet-irreducible $\tilde{p}$. The map $\tilde{}$ then defines a bijective correspondence between join-irreducibles and meet-irreducibles. The poset of join-irreducibles of the computational equivalence lattice $\mathrm{D}/\square_f$ is isomorphic as an ordered set to the subset of D comprising P_f and the set of elements $\{\tilde{q} \mid q \in Q_f\}$. The elements q and $\tilde{q}$ are related by the characteristic property:

$$(\forall x \in \mathrm{FDL}(n) : x \leq q \ \mathbf{xor} \ x \geq \tilde{q}).$$

If p and q are elements of P_f and Q_f respectively, then $p \leq f \leq q$, showing that p and $\tilde{q}$ are comparable only if $p < \tilde{q}$. If f is an element of $\mathrm{FDL}(n)$ that depends upon fewer than n inputs (cf [4]), then f does not depend upon a free generator x. But then $x \geq \tilde{q}$ necessarily, so that no relation of the form $p \leq \tilde{q}$ pertains. $\mathrm{FDL}(n)/\square_f$ is thus Boolean.

Theorem 2.3 has a simple interpretation. If an element of D is viewed in its relationship to the elements of P_f and Q_f, its computational usefulness is precisely correlated with how closely this relationship approximates that of the element f — the unique element less than all elements q in Q_f and greater than all elements p in P_f. This observation can be applied to $\mathrm{FDL}(n)$ to derive Theorem 2.4 below — a classical interpretation of formulae for monotone Boolean functions, recently rediscovered and analysed in the work of Karchmer et al [16]. Some preliminary definitions are first required:

For $f \in \mathrm{FDL}(n)$, let R_f be the rectangular array whose rows are indexed by the elements of P_f, whose columns are indexed by the elements of Q_f, and whose (p,q)-th entry is the intersection of the sets of indices of generators appearing in p and q. A subset of R_f comprising the elements that lie in selected sets of rows and columns, not necessarily contiguous, is a *generalised rectangle*. Two generalised rectangles are *conformal* if they have representatives in exactly the same set of rows or exactly the same set of columns.

Theorem 2.4 *Suppose that $f \in FDL(n)$. Then*

(i) *A monotone formula for f corresponds to a partition of R_f into blocks, each of which is a generalised subrectangle in which all entries have an index in common, having the property that R_f can be built up from its constituent blocks by repeatedly amalgamating conformal pairs of blocks. The number of blocks in the partition is the number of inputs to the formula.*

(ii) *A monotone circuit for f corresponds to a sequence of operations performed on R_f, each of which introduces a new index into R_f, such that on termination all entries in R_f have an index in common. Each operation is defined by selecting a pair of indices (x,y); wherever x and y appear in the same row (respectively column) of R_f, a new index is adjoined to every entry containing x or y in that row (respectively column). The size of the circuit is the length of the sequence.*

In both constructions, row and column operations respectively correspond to AND and OR gates.

Proof :   (i) A monotone formula for the Boolean function f can be interpreted as a circuit in the form of a tree, rooted at the output gate, whose leaves are labelled by input variables, and whose internal nodes correspond to AND and OR gates. To each node ν of this tree, attach a code $< \alpha, \beta >$, where α (resp. β) is a binary sequence indexed by P_f (resp. Q_f) such that α_p (resp. β_q) is 1 if and only if the function computed by ν has p as an implicant (resp. q as a clause).

To derive a partition of R_f, a simple procedure is first applied to each node in the tree on a top-down, breadth-first basis. The effect is in general to alter the codes for all nodes other than the root in a systematic fashion. Suppose that all ancestors in the tree of the node corresponding to the gate ν have been processed, and that ν is an AND gate with the modified code $< \alpha', \beta' >$. (The dual case, when ν is an OR gate, is left to the reader.) All 1's in the binary sequence α' must be common to the α-component of the codes of both gates that are direct inputs to ν; the codes of both these input gates are altered to take the form $< \alpha', * >$. Each 1 that appears in β' must appear in the β-component of the code of at least one of the input gates to ν. Select a representative for each such 1 from one of the gates only, and modify the codes for the input gates to the form $< \alpha', \beta'' >$, where the only 1's in β'' are representatives for 1's in ν.

The blocks of the required partition are then defined by the generalised rectangles in R_f whose rows and columns are respectively indexed by α^ι and β^ι, where $< \alpha^\iota, \beta^\iota >$ is the code allocated to the leaf node associated with input gate ι.

It remains to show that every partition of the stated form is associated with a formula. To this end, observe that the process of representing R_f as an amalgamation of blocks can be represented by a tree whose leaves correspond to the blocks of the partition, whose internal nodes represent row and column operations that amalgamate generalised rectangles, and whose root represents the operation constructing the array R_f. Let each node ν of the tree be labelled by a code $< \alpha, \beta >$, where α (resp. β) is a binary sequence indexed by P_f (resp. Q_f) such that α_p (resp. β_q) is 1 if and only if the generalised rectangle associated with ν intersects the row indexed by p (resp. column indexed by q). Assume that there are N blocks in the partition, and let $< \alpha^\iota, \beta^\iota >$ be the code associated with the ι-th block for $1 \leq \iota \leq N$. By Theorem 2.3, by substituting AND and OR gates for nodes that respectively represent row and column operations, the tree can be re-interpreted as a monotone circuit that computes f from any set of inputs $z_1, z_2, \ldots, z_n$ with the property that $z_\iota \geq p$ whenever $\alpha^\iota_p = 1$ and $z_\iota \leq q$ whenever $\beta^\iota_q = 1$. By hypothesis, for $1 \leq \iota \leq N$, all entries in the ι-th block of the partition have an index $i(\iota)$ in common. It follows that, on replacing the input z_ι by $x_{i(\iota)}$ for $1 \leq \iota \leq N$, the circuit is a monotone formula expressing f in terms of the standard inputs $x_1, x_2, \ldots, x_n$.

(ii) Consider a monotone Boolean circuit Γ that computes f. If g is the output of a node of Γ, the computational equivalence class of g relative to f is entirely determined by its relationship to the prime clauses and prime implicants of f. This equivalence class will be represented in R_f by an index that appears in the (p, q)-th position whenever $p \leq g \leq q$. With this convention, the n inputs are represented by the indices $1, 2, \ldots, n$ in the array R_f in its initial form. The circuit Γ can be represented as an acyclic digraph with edges directed from the set of inputs to the single output node. If the nodes of Γ are enumerated in a manner consistent with their dependency, it is possible to introduce indices systematically to represent the associated computational equivalence classes in R_f. For instance, let g be the output of an AND gate whose direct inputs compute h and k. If p and q respectively represent a prime clause and a prime implicant of f, then $p \leq g$ iff $p \leq h$ and $p \leq k$, whilst $g \leq q$ iff $h \leq q$ or $k \leq q$. Introducing g then corresponds to adjoining the new index representing g to each entry in R_f in which at least one of h or k is represented, and such that both h and k are represented within the corresponding row. The function f is uniquely represented by an index that appears in all entries of the array R_f. In this way, the circuit Γ can be represented by a sequence of operations on R_f as described in the Theorem.

The reverse process of constructing a circuit from a sequence of operations on R_f is a straightforward application of Theorem 2.3. Each index m can be interpreted as representing a gate that computes a function g with the property that $g \geq p$ (resp. $g \leq q$) whenever index m appears in the row indexed by p (resp. the column indexed by q). In this way, the occurrences of an index in the array R_f are a geometrical representation of the computational equivalence class modulo f to which the corresponding gate belongs. The introduction of a new index z on selection of indices x and y then represents in geometric terms the effect of introducing a gate whose inputs are the gates associated with x and y.   □

Theorem 2.4 indicates a strong connection between configurations of prime implicants and clauses and the computational characteristics of a monotone Boolean function — a theme that has been developed in the work of Wigderson and others. Wigderson and Karchmer have applied an information-theoretic analysis to this representation to derive significant lower bounds [17]. Karchmer et al. [16] have shown that the array R_f consists exclusively of singletons if and only if f can be expressed as a formula without duplication of the inputs — a result that has been independently derived by others, including Paterson and the author [14, 19, 20]. Razborov has also derived lower bounds by considering alternative representations of the matrix R_f [21]. Analysis of the structure of R_f in its entirety is technically difficult. As Razborov has demonstrated, there are also limitations on the quality of the lower bounds that can be obtained by considering standard invariants of matrices alone. In this paper, generic substructures common to all arrays of the form R_f are identified this may lead eventually to alternative methods for their analysis. These substructures stem from an alternative model for FDL(n), based on the concept of a combinatorially-piecewise linear map.

3.   An alternative model for free distributive lattices

This section considers an alternative combinatorial model for finite free distributive lattices. This is defined in terms of combinatorially piecewise linear (cpl) maps, as first introduced by the author in [3]. The original motivation for introducing cpl maps was the proof of a geometric theorem concerning the representability of piecewise linear functions as pointwise maxima of minima of linear functions. In this section, the elementary properties of cpl maps are outlined. For more background details, the interested reader may consult [3, 4, 5].

3.1.   Characteristics of the combinatorial model

3.1.1.   Combinatorially piecewise-linear maps

Let Σ_n denote the symmetric group of permutations of $\{1, 2, \ldots, n\}$. Σ_n has a standard presentation relative to the generating set consisting of the $n - 1$ transpositions $\tau_1, \tau_2, \ldots, \tau_{n-1}$, where τ_i interchanges i and $i + 1$. The explicit set of relations which defines this presentation is:

$$
\begin{aligned}
\tau_i^2 &= 1 & (1 \leq i \leq n - 1) \\
(\tau_i \tau_{i+1})^3 &= 1 & (1 \leq i \leq n - 2) \\
(\tau_i \tau_j)^2 &= 1 & (2 \leq i + 1 < j \leq n - 1).
\end{aligned}
$$

The Cayley diagram associated with this presentation is then an $(n\text{-}1)$-edge-coloured combinatorial graph Γ_n in which the nodes are in 1-1 correspondence with permutations in Σ_n and there is a bi-directed edge of colour i from ρ to σ if and only if $\sigma = \tau_i \rho$ (where '$\tau_i \rho$' denotes 'τ_i followed by ρ'). Following the conventions defined in [3], the subset $\mathcal{S}$ of $\{1, 2, \ldots, n\}$ is said to be a *vertex* of the permutation σ in Σ_n if $\mathcal{S} = \{\sigma(1), \sigma(2), \ldots, \sigma(|\mathcal{S}|)\}$.

A map $\Sigma_n \rightarrow \{1, 2, \ldots, n\}$ is *combinatorially piecewise-linear* (of degree n) if it corresponds to an n-colouring of Γ_n in which nodes ρ and σ that are adjacent via an i-coloured edge either have the same colour or have colours selected from the set $\{\sigma(i), \sigma(i+1)\} = \{\rho(i), \rho(i+1)\}$. This class of maps was introduced in [3] in connection with the study of representation of piecewise-linear functions as pointwise maxima of minima of linear functions. There is a 1-1 correspondence between cpl maps of degree n and elements of FDL(n), the finite distributive lattice freely generated by n generators $x_1, x_2, \ldots, x_n$. In one direction, this correspondence is defined by associating with the cpl map F the function from the lattice of subsets of $\{1, 2, \ldots, n\}$ to $\{0, 1\}$ which maps $\mathcal{S}$ to 1 if and only if $F(\sigma) \in \mathcal{S}$ whenever $\mathcal{S}$ is a vertex of σ (cf [3]). The inverse is then defined by mapping the element f in FDL(n) to the cpl map F such that $F(\sigma) = \sigma(r)$, where $f(\{\sigma(1), \sigma(2), \ldots, \sigma(r - 1)\}) = 0$ and $f(\{\sigma(1), \sigma(2), \ldots, \sigma(r)\}) = 1$. Figure 3.1 depicts the 4-colouring of the Cayley diagram Γ_4 associated with the element $(x_1 \wedge x_2) \vee (x_2 \wedge x_3 \wedge x_4) \vee (x_1 \wedge x_4)$ in FDL(4). For instance, the value of the cpl map on the permutation with image representation 3214 is 1 — the index underlined at the node representing this permutation.

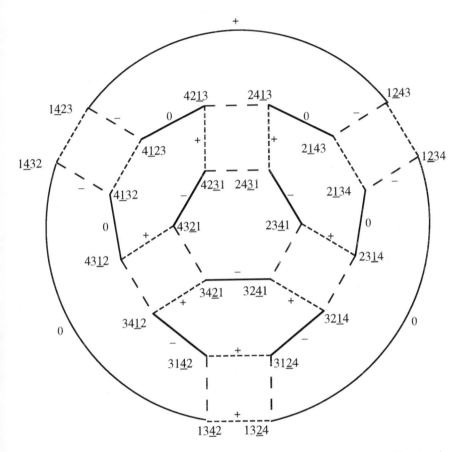

Figure 3.1. The cpl map associated with $(x_1 \wedge x_2) \vee (x_2 \wedge x_3 \wedge x_4) \vee (x_1 \wedge x_4)$.

Given a cpl map F, an edge $(\sigma, \sigma' = \tau_r \sigma)$ is *singular* for F if $F(\sigma)$ — and necessarily also $F(\sigma')$ — is in the set $\{\sigma(r), \sigma(r+1)\}$. Three types of singular edge can be distinguished:

- a *positive* singularity, when $F(\sigma) = \sigma(r)$ and $F(\sigma') = \sigma(r+1)$, or equivalently $F(\sigma') = \sigma'(r)$ and $F(\sigma) = \sigma'(r+1)$,

- a *negative* singularity, when $F(\sigma) = \sigma(r+1)$ and $F(\sigma') = \sigma(r)$, or equivalently $F(\sigma') = \sigma'(r+1)$ and $F(\sigma) = \sigma'(r)$,

- a *zero* singularity, when $F(\sigma') = F(\sigma)$.

If F is a cpl map such that $F(\sigma) = \sigma(r)$, then r is the *rank* of F at σ. The positive and negative singular edges for F can then be characterised as those edges across which F changes in value but not rank, and the zero singular edges as those across which F changes in rank but not in value. A duality for cpl maps gives another perspective on this concept of singularity.

3.1.2. Dual cpl maps

There are two Cayley diagrams associated with the presentation of Σ_n defined above; Γ_n in which adjacency of nodes is defined by multiplication by a transposition τ_i on the left, and a dual graph $\Gamma_n^\star$, in which adjacency is defined via multiplication by a transposition τ_i on the right. (Note that different pairs of permutations are adjacent in Γ_n and $\Gamma_n^\star$.) A map $\Sigma_n \to \{1, 2, \ldots, n\}$ is *dual combinatorially piecewise-linear* (of degree n) if it corresponds to an n-vertex-colouring of $\Gamma_n^\star$ in which nodes ρ and σ that are adjacent via an i-coloured edge either have the same colour or have colours selected from the set $\{i, i+1\}$. As is the case for cpl maps, there is a 1-1 correspondence between elements of $\mathrm{FDL}(n)$ and dual cpl maps. Given $f \in \mathrm{FDL}(n)$, there is an associated dual cpl map $F^\star : \Sigma_n \to \{1, 2, \ldots, n\}$ defined as follows. Let T be the total ordering of $\{1, 2, \ldots, n\}$ defined by $1 > 2 > \ldots > n$, and let $F^\star(\sigma)$ be the image of $f \in \mathrm{FDL}(n)$ under the canonical lattice morphism $\Pi_\sigma : \mathrm{FDL}(n) \to T$ mapping the generator x_i of $\mathrm{FDL}(n)$ to $\sigma(i)$. Then:

Lemma 3.1 $F^\star$ *is a dual cpl map. Moreover, if F is the cpl map associated with $f \in \mathrm{FDL}(n)$, then $F^\star(\sigma) = \sigma(F(\sigma^{-1}))$ and $F(\sigma) = \sigma(F^\star(\sigma^{-1}))$.*

Figure 3.2 depicts the dual cpl map associated with the element

$$(x_1 \wedge x_2) \vee (x_2 \wedge x_3 \wedge x_4) \vee (x_1 \wedge x_4)$$

in $\mathrm{FDL}(4)$, which can be compared with the corresponding cpl map in Figure 3.1. In pictorial terms, Lemma 3.1 asserts that if Figures 3.1 and 3.2 are superimposed, then at each node the value of the function specified in the one diagram is the rank of the function specified in the other.

3.1.3. Chains and cycles of singular edges

Theorem 3.2 *Let f be a cpl map $\Gamma_n \to \{1, 2, \ldots, n\}$. Then f has two singular edges at those nodes σ for which $1 < rk(f(\sigma)) < n$, and one singular edge at those nodes σ for which $rk(f(\sigma))$ is 1 or n. The singular edges of f can be represented as a disjoint union of simple circuits exclusively visiting nodes σ at which $1 < rk(f(\sigma)) < n$, and simple paths that originate and terminate at nodes σ such that $rk(f(\sigma)) = 1$ or $rk(f(\sigma)) = n$.*

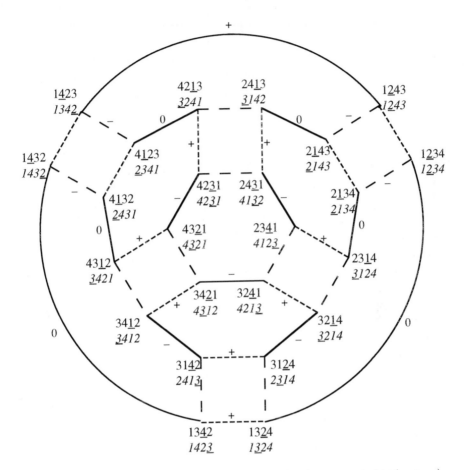

Figure 3.2. The cpl map associated with $(x_1 \wedge x_2) \vee (x_2 \wedge x_3 \wedge x_4) \vee (x_1 \wedge x_4)$ with the corresponding dual cpl map superimposed.

The paths and circuits in Theorem 3.2 are respectively called the singular chains and singular cycles of f. Apart from the trivial cpl maps defined by the free generators of FDL(n), there are no cpl maps for which there simultaneously exist permutations σ and ρ such that $rk(f(\sigma)) = 1$ and $rk(f(\rho)) = n$. Elements of FDL(n) incomparable with the free generators have only singular cycles, and these define a partition of Σ_n. For any element of FDL(n), the minimal (resp. maximal) rank of the associated cpl map is determined by the size of the shortest prime implicant (resp. clause). Theorem 3.2 shows that a cpl map is characterised by its set of singular chains and cycles.

3.2.   Cycles of singular edges as relations in Σ_n

The analysis of singular chains and cycles can be approached from both group- and lattice-theoretic perspectives. This section focuses on singular cycles as relations between the transpositions that generate Σ_n. Generic properties of the relations are identified, and a useful method of representation is introduced.

Every cycle of singular edges corresponds to a relation in Σ_n defined by a product of transpositions in which each consecutive pair has adjacent indices. For instance, the element $(x_1 \wedge x_2) \vee (x_2 \wedge x_3 \wedge x_4) \vee (x_1 \wedge x_4)$ in $\mathrm{FDL}(4)$ gives rise to a cpl map having two cycles, one of length 6 corresponding to the relation:

$$\tau_1 \tau_2 \tau_1 \tau_2 \tau_1 \tau_2 = \left(\tau_1 \tau_2\right)^3 = 1, \tag{1}$$

and one of length 18 corresponding to the relation:

$$\tau_1 \tau_2 \tau_3 \tau_2 \tau_1 \tau_2 \tau_3 \tau_2 \tau_3 \tau_2 \tau_1 \tau_2 \tau_1 \tau_2 \tau_3 \tau_2 \tau_3 \tau_2 = 1. \tag{2}$$

It can be shown that the relations which correspond to singular cycles have a special form: they can be expressed as concatenations of products of transpositions in which increasing and decreasing sequences of indices alternate. Explicitly, with each increasing chain of indices

$$a < b < c < \ldots < x < y < z \tag{3}$$

there is an associated product of transpositions

$$\tau_a \tau_{a+1} \cdots \tau_z \tau_{z-1} \cdots \tau_b \tau_{b+1} \cdots \tau_y \tau_{y-1} \cdots \tau_c \tau_{c+1} \cdots \tau_x \tau_{x-1} \cdots \cdots \tau_m \tag{4}$$

which can be expressed in the form $\tau_a \pi \tau_m$. The relations associated with singular cycles are then concatenations of alternating sequences of the form $\tau_a \tau_m \pi^R$, where $\tau_a \pi \tau_m$ has the form (4), and π^R denotes the reversal of the segment π. The special notation

$$[a\ z\ b\ y\ c\ x\ \ldots\ ]$$

is adopted for the alternating sequence derived from the chain (3) in this fashion. Using this notation, the specimen relations (1) and (2) defined above can be represented as

$$[1\ 2]^3 \text{ and } [1\ 3]\ [1\ 3\ 2]\ [1\ 2]\ [1\ 3\ 2]$$

respectively. Two characteristics of cycles are helpful in classification: the *length* of a cycle is the sum of the lengths of its constituent alternating sequences, where

$$\text{length of } [a_1 \ a_2 \ \dots \ a_n] = 2 \cdot \textstyle\sum_{i=1}^{n-1} \mid a_i - a_{i+1} \mid,$$

and the *alternation index* is the number of local maxima, to which each constituent alternating sequence $[a_1 \ a_2 \ \dots \ a_n]$ contributes $n-1$. A cycle of length ℓ with alternation index α will be described as an ℓ_α cycle. (The specimen cycles (1) and (2) above are 6_3 and 18_6 cycles respectively.)

Necessary and sufficient conditions for a relation in Γ_n to correspond to a singular cycle have yet to be derived. By duality, any such cyclic relation must be expressible both as a concatenation of alternating sequences, and as a concatenation of dual alternating sequences, defined in a manner very similar to that described above from a decreasing chain of indices

$$a > b > c > \dots > x > y > z.$$

(Representations of this form for the specimen cycles (1) and (2) above are

$$[2\ 1]^3 \text{ and } [3\ 1][3\ 2][3\ 1\ 2][3\ 2][3\ 1]$$

respectively.) This restriction on relations — the WM-condition — eliminates many possible products of transpositions from consideration, but is not sufficient to characterise those associated with singular cycles. The relation $([1\ 3][1\ 3\ 2])^2$, for instance, traces a circuit of Γ_4 which is not simple — it incorporates the 12_4 cycle $[1\ 3\ 2]^2$. It is of incidental interest to note that singular cycles are not necessarily palindromic: cf the 36_9 cycle

$$[1\ 4\ 2][1\ 4\ 2\ 3][1\ 4\ 3][1\ 3\ 2].$$

In general, a singular cycle in Σ_n has a representation α as a product of alternating sequences in which the set of indices of transpositions represented is $\{a_1, a_2, \dots, a_k\}$ where $1 \le a_1 < a_2 < \dots < a_k < n$. For reasons to be explained in §3.3, the product of alternating sequences obtained from α by replacing all occurrences of a_i in α by i, for $1 \le i \le k$, also represents a singular cycle, defined in Σ_{k+1}. Singular cycles whose representations involve the set of indices $\{a_1, a_2, \dots, a_k\}$ rather than $\{1, 2, \dots, k\}$ are essentially variants of smaller cycles, associated with cpl maps of lower degree. For instance, the relation $[1\ 3]^3 = 1$, which gives rise to a singular cycle of length 12 having three maxima is a variant of the 6_3 cycle $[1\ 2]^3$ into which six zero singular edges have been interpolated. In classifying singular cycles, it then suffices to consider only those cycles whose representations involve the indices $\{1, 2, \dots, k\}$. For each k, all such cycles arise from cpl maps of degree $k+1$.

A catalogue of relations known to arise as singular cycles in this way appears as an Appendix to [5]. Each relation has a dual obtained by expressing the relation as a concatenation of dual alternating sequences, and relabelling the indices $1, 2, \ldots, n$ by $n, n-1, \ldots, 1$. Under this duality the cycle 6_3 is self-dual, whilst the 18_6 cycle

$$[1\ 3][1\ 3\ 2][1\ 2][1\ 3\ 2] = [3\ 1][3\ 2][3\ 1\ 2][3\ 2][3\ 1]$$

has as dual $[1\ 3][1\ 2][1\ 3\ 2][1\ 2][1\ 3]$, which is associated with the dual of $(x_1 \wedge x_2) \vee (x_2 \wedge x_3 \wedge x_4) \vee (x_1 \wedge x_4)$, viz: $(x_1 \vee x_2) \wedge (x_2 \vee x_4) \wedge (x_1 \vee x_4) \wedge (x_1 \vee x_3)$. A cycle is not characterised up to duality by its length and alternation index — for instance, $([1\ 4\ 2][1\ 3])^2$ and $[1\ 4\ 3]\ [1\ 5\ 2\ 4\ 3]$ are distinct 28_6 cycles.

Non-trivial relations arising as singular cycles of elements of FDL(4) are:

12_4	:	$[1\ 3\ 2]^2$	dual of $([1\ 3][1\ 2])^2$
18_6	:	$[1\ 3][1\ 3\ 2][1\ 2][1\ 3\ 2]$	dual of $[1\ 2][1\ 3]^2[1\ 2][1\ 3\ 2]$
24_8	:	$([1\ 3\ 2][1\ 2][1\ 3])^2$	self-dual
24_9	:	$([1\ 2][1\ 3\ 2])^3$	self-dual

(It can be shown that no cpl map of degree higher than 4 has a Hamiltonian circuit as a singular cycle [5].)

3.3.   Singular cycles and intervals in FDL(n)

This section adopts a lattice-theoretic approach to the classification of singular cycles. It examines the conditions under which a particular monotone Boolean function f gives rise to a cpl map possessing a particular cycle.

Certain characteristic properties of the class of monotone Boolean functions that give rise to a cpl map with a particular singular cycle $\mathcal{C}$ can be inferred from its representation as a product of alternating sequences. Suppose that f is a monotone Boolean function in FDL(n) that gives rise to the cpl map F with $\mathcal{C}$ as a singular cycle. Every local maximum of $\mathcal{C}$ corresponds to a positive singularity of F, every local minimum corresponds to a negative singularity and all other edges are zero singularities. If the representation for the singular cycle $\mathcal{C}$ involves exactly k distinct indices, the behaviour of F on $\mathcal{C}$ coincides with that of a cpl map F' derived from a restriction f' obtained from f by assigning values to all but $k+1$ free variables. $\mathcal{C}$ is then a variant of a singular cycle associated with the cpl map of degree $k+1$ that corresponds to f' in FDL($k+1$). The nature of the restriction process can be understood by examining the behaviour of F on $\mathcal{C}$ more closely.

The singular cycle $\mathcal{C}$ is a subset of Σ_n. As in §3.1.1, if σ is a permutation in $\mathcal{C}$, then $F(\sigma) = \sigma(r)$, where $f(\{\sigma(1), \sigma(2), \ldots, \sigma(r-1)\}) = 0$ and $f(\{\sigma(1), \sigma(2), \ldots, \sigma(r)\}) = 1$. For brevity, it will be convenient to refer to a monotone Boolean function g that gives rise to a cpl map G with the same behaviour as F on $\mathcal{C}$ as *compatible* with F on $\mathcal{C}$. As σ ranges over $\mathcal{C}$, the rank r of $F(\sigma)$ ranges over an interval $[s,t]$, where $1 < s < t < n$. The vertex $\{\sigma(1), \sigma(2), \ldots, \sigma(s-1)\}$ is necessarily common to all permutations in $\mathcal{C}$, and the definition of F above shows that the function f' derived from f by assigning free variables with indices from the set $\{\sigma(1), \sigma(2), \ldots, \sigma(s-1)\}$ to 0 is compatible with F on $\mathcal{C}$. By a dual argument, the further restriction derived from f' by assigning free variables with indices from the set $\{\sigma(t+1), \sigma(t+2), \ldots, \sigma(n)\}$ to 1 is also compatible. To complete the restriction process, consider the set of values $F(\mathcal{C})$ attained by the cpl map F on $\mathcal{C}$. By hypothesis, $F(\mathcal{C})$ is a subset of $\{\sigma(s), \sigma(s+1), \ldots, \sigma(t)\}$; if it is a proper subset, then further restriction of f through assignment of arbitrary 0 and 1 values to the free variables with indices in $\{\sigma(s), \sigma(s+1), \ldots, \sigma(t)\} \setminus F(\mathcal{C})$ leads once more to a function compatible with F on $\mathcal{C}$.

The general principles described above will be illustrated by considering a particular example. The cpl map associated with the element

$$f = (x_1 \wedge x_2) \vee (x_2 \wedge x_4) \vee (x_1 \wedge x_4) \vee (x_1 \wedge x_3 \wedge x_5)$$

in FDL(5) has a set of singular cycles that can be classified as follows:

- Singular cycles which incorporate a permutation ending in 3 or 5:
  All permutations in such cycles have the same final index. The behaviour of f on such cycles is therefore the same as that of the threshold function $(x_1 \wedge x_2) \vee (x_2 \wedge x_4) \vee (x_1 \wedge x_4)$ that results from setting the variable x_3 or x_5 to 0. The permutations 12435, 31245, 12453 and 51243 in which the indices 1, 2 and 4 appear contiguously lie on distinct singular 6_3 cycles, and the permutations 12345 and 12543 lie on distinct singular 12_3 cycles which are variants of the 6_3 cycles.

- Singular cycles which contain a permutation beginning with 3 or 5:
  All permutations in such cycles have the same initial index. The behaviour of f on such cycles is the same (up to symmetry) as that of the function $(x_1 \wedge x_2) \vee (x_2 \wedge x_4) \vee (x_1 \wedge x_4) \vee (x_1 \wedge x_3)$ that results from setting the variable x_5 to 1. The permutation 51234 lies on an 18_6 cycle: [1 3][1 2][1 3 2][1 2][1 3], as also (by symmetry) does 31254.

- All permutations not beginning or ending in 3 or 5 form a singular 36_8 cycle.

The lattice-theoretic significance of possessing a particular singular cycle can also be related to the array R_f associated with the element f in FDL(n) as in Theorem 2.4 above. For non-trivial functions f, analysis of the array R_f is difficult, but it can be shown that the presence of particular singular cycles is connected with features of R_f. Let $\Phi(\mathcal{C})$ denote the set of functions $f \in$ FDL(n) that give rise to a cpl map possessing the singular cycle $\mathcal{C}$. If

Cycle $\mathcal{C}$	1	2	3	2	3	2	1	2	3	4	3	2	3	2	3	4	3	2
$Cl(\mathcal{C})$			145		135					25			234			25		
Cpl map	5	5	5	5	5	5	5	5	5	5	2	2	2	2	2	2	5	5	5
	4	4	4	1	1	3	3	3	3	2	5	3	3	4	4	5	2	4	4
	3	3	1	4	3	1	4	4	2	3	3	5	4	3	5	4	4	2	3
	2	1	3	3	4	4	1	2	4	4	4	4	5	5	3	3	3	3	2
	1	2	2	2	2	2	2	1	1	1	1	1	1	1	1	1	1	1	1
$Imp(\mathcal{C})$		12			234			12						145			135	

Figure 3.3. Implicants and clauses that define a singular cycle

$\Phi(\mathcal{C})$ contains f and g such that $f \leq g$ in $\mathrm{FDL}(n)$, then it also contains the interval $[f,g]$. Moreover, $\Phi(\mathcal{C})$ is closed with respect to $\vee$ and $\wedge$. This means that $\Phi(\mathcal{C})$ is an interval in $\mathrm{FDL}(n)$, or equivalently, that a necessary and sufficient condition for $f \in \Phi(\mathcal{C})$ is that f possesses a particular set of implicants and clauses. Each set of implicants and clauses that arises in this way is associated with a characteristic pattern of intersection between prime implicants and prime clauses in the array R_f.

To illustrate this, consider the conditions under which an element f in $\mathrm{FDL}(5)$ gives rise to a cpl map with the 18_5 singular cycle $[1\ 3\ 2][1\ 4\ 2\ 3]$ at the permutation 12345. Tracing this singular cycle $\mathcal{C}$ and monitoring the implicants and clauses required to ensure the appropriate behaviour, as depicted in Figure 3.3, shows that f has the set of implicants $Imp(\mathcal{C})$:

$$\{x_1 \wedge x_2,\ x_2 \wedge x_3 \wedge x_4,\ x_1 \wedge x_3 \wedge x_5,\ x_1 \wedge x_4 \wedge x_5\}$$

— associated with negative singularities on $\mathcal{C}$, and the set of clauses $Cl(\mathcal{C})$:

$$\{x_1 \vee x_4 \vee x_5,\ x_1 \vee x_3 \vee x_5,\ x_2 \vee x_5,\ x_2 \vee x_3 \vee x_4\}$$

— associated with positive singularities on $\mathcal{C}$.

Conversely, if the monotone Boolean function g has $Imp(\mathcal{C})$ as a subset of its implicants and $Cl(\mathcal{C})$ as a subset of its clauses, its restriction to the cycle $\mathcal{C}$ is necessarily as depicted in Figure 3.3. To prove this, observe that the implicants in $Imp(\mathcal{C})$ (resp. clauses in $Cl(\mathcal{C})$) ensure that g agrees with f at the endpoints of its negative (resp. positive) singular edges in $\mathcal{C}$. If σ is a permutation on $\mathcal{C}$ that is adjacent to zero singular edges of f in $\mathcal{C}$, then there is a nearest singular edge of f that is not zero singular on each side of σ on the cycle C. Without loss of generality, suppose that the nearest such singular edges e_l and e_r to the left and right of σ on C are respectively negative and positive. If p is the implicant (resp. q is the clause) associated with e_l (resp. e_r), then it is clear from Figure 3.3 that p and q have one variable in common, and that the index of this variable is the value of f at σ, say $\sigma(r)$. But the indices of variables in p are a subset of the set $\{\sigma(1), \sigma(2), \ldots, \sigma(r)\}$, and the indices of variables in q are a subset of $\{\sigma(r), \sigma(r+1), \ldots, \sigma(n)\}$, from which it follows that g agrees with f at σ.

The above discussion shows that the behaviour of a monotone Boolean function f on a singular cycle is determined by pairs (p, q) consisting of an implicant p and a clause q of f having a single variable in common. Such pairs are in turn associated with characteristic patterns of intersection in the array R_f. In the context of Figure 3.3, the pattern of intersection between these implicants and clauses of f defines the 4×4 array:

$$
\begin{array}{cccc}
1 & 1 & 2 & 2 \\
4 & 3 & 2 & 234 \\
15 & 135 & 5 & 3 \\
145 & 15 & 5 & 4
\end{array}
$$

This pattern does not itself correspond directly to a pattern of intersection between prime implicants and prime clauses of any particular function f giving rise to the singular cycle C, though it can differ only in respect of non-singleton entries. An appropriate way to describe the possible configurations of intersections associated with such elements f is to extract from each row (respectively column) any indices which do not appear as singletons, and to record these separately in an additional column (respectively row). These extra row (respectively column) entries correspond to variables that may or may not appear in the implicant (respectively clause) associated with the row (respectively column) — depending on the choice of f. Since the form of the non-singleton entries in the array is dependent on the choice of function f, and is uniquely determined by which of the optional variables appear in prime implicants and prime clauses of f, it is then convenient to omit these entries from the array. With this convention, the pattern of intersection that is associated with the 18_5 singular cycle $[1\ 3\ 2][1\ 4\ 2\ 3]$ at the permutation 12345 is represented by the *configuration*:

$$
\begin{array}{ccccc}
 & 1 & 1 & 2 & 2 \\
 & 4 & 3 & 2 & * \\
1 & * & * & 5 & 3 \\
1 & * & * & 5 & 4 \\
 & 5 & 5 & &
\end{array}
$$

The classification of configurations is a variant of the problem of classifying intervals in $FDL(n)$ associated with singular cycles. It is easy to see that any array that defines such a configuration contains at least two distinct indices in every row and column. Restriction of a function by setting an input x_i appearing within a configuration to 0 (respectively 1) can be interpreted as 'deleting all rows (respectively columns) containing index i'. If the effect of such an operation is to produce a subconfiguration, i.e., an array with at least two distinct indices in every row and column, this indicates the presence of additional singular cycles. For instance, the 36_8 cycle $([1\ 3\ 2][1\ 4]^2)^2$ has the characteristic configuration:

$$
\begin{array}{cccc}
2 & 2 & 1 & * \\
4 & 4 & * & 1 \\
* & * & 4 & 2 \\
3 & 5 & 1 & 1
\end{array}
$$

This configuration is uniquely associated with the monotone Boolean function in $FDL(5)$ that has the implicants $x_1 \wedge x_2, x_2 \wedge x_4, x_1 \wedge x_4$ and $x_1 \wedge x_3 \wedge x_5$ as its prime implicants, and the clauses $x_1 \vee x_2, x_1 \vee x_4, x_2 \vee x_3 \vee x_4$ and $x_2 \vee x_4 \vee x_5$ as its prime clauses. The configuration yields subconfigurations on various combinations of assignments from the set $\{x_3 = 0, x_5 = 0, x_3 = 1, x_5 = 1\}$. In this way, it is possible to derive the complete classification of singular cycles for $f = (x_1 \wedge x_2) \vee (x_2 \wedge x_4) \vee (x_1 \wedge x_4) \vee (x_1 \wedge x_3 \wedge x_5)$ above.

4.   Conclusions and directions for further work

The work described in this paper forms part of a broad research programme aimed at finding combinatorial representations for monotone Boolean functions that are more informative where complexity is concerned. Representations that can aid intuition would be particularly helpful if current methods of determining complexity bounds on Boolean functions are to be improved. Links with combinatorial geometry, such as are suggested by Theorem 2.2(iii) and the related geometric models underlying cpl maps, may offer good prospects.

The role that configurations can play in the analysis of arrays of the form R_f cannot be evaluated at this stage. Further research is required to determine the nature and significance of singular cycles. It seems plausible that the relations between alternating sequences which correspond to singular cycles are precisely those which satisfy the WM-condition and define simple circuits of Γ_n. Bounds on the length and alternation index for singular cycles would also be of interest. It is relatively easy to exhibit generic families of cycles whose length grows quadratically (cf [5]), but no asymptotic upper bounds on length have been found. The partition of Γ_n associated with f also suggests the possible existence of a non-standard canonical decomposition for f.

To date, the most effective application of the ideas developed in this paper is that described in [6], where the class of monotone Boolean functions computable by planar monotone circuits is characterised; some refinements and extensions appear in [13]. A significant feature of this characterisation is the use made of the general form of Theorem 2.3 as it applies not simply to FDL(n) but also to its algebraic quotients. It is also interesting as an illustration of how algebraic considerations can be used to construct circuits directly.

A more unexpected application of Theorem 2.3, to be described in [1], relies upon relating planar comparator circuits for sorting and simple arrangements of pseudo-lines in the plane. This relationship is derived via the association between geometry and circular sequences (cf [18]). By applying Theorem 2.3, it can be shown that any simple arrangement of n pseudo-lines defines at least $n-2$ minimal regions that are triangular. Future research will be directed towards finding similar geometric representations for the circular sequences that arise as singular cycles. The study of the weak ordering of Coxeter groups provides a general setting for this research [9].

5.   Acknowledgments

I am indebted to Mike Paterson for helpful discussions and motivating ideas. I am also grateful to an anonymous referee for several helpful suggestions.

References

[1] Atkinson, M.D., Beynon, W.M., *A note on the combinatorics of simple arrangements. (in preparation)*

[2] Berkowitz, S.J., *On some relationships between monotone and non-monotone complexity.* Tech. Report, Comput. Sci. Dept., Univ. of Toronto, 1982.

[3] Beynon, W.M., *Combinatorial aspects of piecewise linear maps.* J. London Math. Soc. (2) 7, 1974, pp. 719-727.

[4] Beynon, W.M., *Replaceability and computational equivalence for monotone Boolean functions.* Acta Informatica 22, 1985, 433-449.

[5] Beynon, W.M., *Monotone Boolean functions as combinatorially piecewise linear maps.* Dept. of Computer Science Research Report 109, Univ. of Warwick, 1987.

[6] Beynon, W.M., Buckle, J.F., *On the planar monotone computation of Boolean functions*. Theoretical Computer Science 53, 1987, 267-279.

[7] Beynon, W.M., Buckle, J.F., *Computational equivalence and replaceability in finite algebras*. Theory of Computation Report 72, Univ. of Warwick, 1985.

[8] Birkhoff, G., *Lattice Theory*. AMS. Colloq. Pubs. XXV, 3rd ed., AMS RI, 1967.

[9] Bjørner, A., *Orderings of Coxeter groups*. Proc. 'Combinatorics and Algebra', Boulder, June, 1983, Amer. Math. Soc. 1983.

[10] Buckle, J.F., *Computational Aspects of Lattice Theory*. PhD thesis, Univ. of Warwick, 1990.

[11] Dunne, P.E., *The complexity of Boolean networks*. Academic Press, 1988.

[12] Edelsbrunner, H., *Algorithms in Computational Geometry*. Springer-Verlag, 1987.

[13] Fischer, P., *Which Boolean functions can be computed by monotone planar circuits*. Forschungsbericht Nr. 300, Universität Dortmund, 1989.

[14] Gurvich, V.A., *On repetition-free Boolean functions*. sl Uspekhi Matematicheskikh Nauk 32(1), 1977, 183-184, (in Russian).

[15] Jurgensen, H., Thierrin, G., *Semigroups with each element disjunctive*. Semigroup Forum Vol. 2, 1980, 127-141.

[16] Karchmer, M., Linial, N., Newman, I., Saks, M., Wigderson, A., *Combinatorial characterization of read-once formulae*. Discrete Math., to appear.

[17] Karchmer, M., Wigderson, A., *Monotone circuits for connectivity require super-logarithmic depth*. Proc. ACM 20th STOC, 1988, 539-550.

[18] Knuth, D.E., *Axioms and Hulls*. Report No. 7, Institut Mittag-Leffler, 1991.

[19] Mundici, D., *Functions computed by monotone Boolean formulas with no repeated variables*. Theoretical Computer Science 66, 1989, 113-114.

[20] Newman, I., *On read-once Boolean functions*. This volume.

[21] Razborov, A.A., *Applications of matrix methods to the theory of lower bounds in computational complexity*. Combinatorica, 10, 1990, 81-93.

[22] Shyr, H.J., *Free monoids and languages*. Lecture Notes, Dept. of Maths., Soochow Univ., Taipei, Taiwan, 1979.

Monotone Complexity

*Michelangelo Grigni**   *Michael Sipser*[†]

Abstract

We give a general complexity classification scheme for monotone computation, including monotone space-bounded and Turing machine models not previously considered. We propose monotone complexity classes including mAC^i, mNC^i, $mLOGCFL$, $mBWBP$, mL, mNL, mP, $mBPP$ and mNP. We define a simple notion of monotone reducibility and exhibit complete problems. This provides a framework for stating existing results and asking new questions.

We show that mNL (monotone nondeterministic log-space) is not closed under complementation, in contrast to Immerman's and Szelepcsényi's nonmonotone result [Imm88, Sze87] that $NL = co\text{-}NL$; this is a simple extension of the monotone circuit depth lower bound of Karchmer and Wigderson [KW90] for st-connectivity.

We also consider $mBWBP$ (monotone bounded width branching programs) and study the question of whether $mBWBP$ is properly contained in mNC^1, motivated by Barrington's result [Bar89] that $BWBP = NC^1$. Although we cannot answer this question, we show two preliminary results: every monotone branching program for majority has size $\Omega(n^2)$ with no width restriction, and no monotone analogue of Barrington's gadget exists.

1. Introduction

A computation is *monotone* if it does not use the negation operation. Monotone circuits and formulas have been studied as restricted models of computation with the goal of developing techniques for the general problem of proving lower bounds.

---

*Department of Computer Science, University of British Columbia, Vancouver, British Columbia, Canada V6T 1Z2. Work completed while this author was at MIT.

†Department of Mathematics, Massachusetts Institute of Technology, Cambridge, Massachusetts 02139. Research supported by NSF grant 8912586-CCR and DARPA contract N00014-89-J-1988.

In this paper we seek to unify the theory of monotone complexity along the lines of Babai, Frankl, and Simon [BFS86] who gave a framework for communication complexity theory. We propose a collection of monotone complexity models paralleling the familiar nonmonotone models. This provides a rich classification system for monotone functions including most monotone circuit classes previously considered, as well as monotone space-bounded complexity classes which have previously received little attention.   This classification gives a language for discussing existing results and for suggesting new problems.

To illustrate our objective, let us consider two of the main results on monotone complexity: 1) Razborov's theorem showing that the clique function is not computable by polynomial size monotone circuits and 2) Karchmer and Wigderson's theorem showing that the st-connectivity function cannot be computed by log-depth monotone circuits. These may be viewed as saying that, with respect to monotone computation, $P \neq NP$, and that $NC^1 \neq NL$.

We ask which complexity class containments carry over to their monotone analogues. Many of the obvious containments carry through, e.g. the simulation of Turing machines by circuits. In particular we consider two recent surprising containment results in space-bounded complexity:

1. Immerman [Imm88] and Szelepcsényi [Sze87] showed that NL is closed under complementation, and

2. Barrington [Bar89] showed that $BWBP$ (bounded width branching programs) contains all of NC^1.

These results are interesting because the simulation techniques do not seem to carry over to the corresponding monotone models.

We show that the first result does not hold in a monotone world, i.e. the monotone complexity class corresponding to NL is not closed under complementation, so the inductive counting technique used by these authors cannot be replaced by a monotone simulation; the proof of this is an extension of the proof of Karchmer and Wigderson [KW90].

We have not resolved the second result, i.e. whether monotone bounded width branching programs may simulate monotone formulas. Nevertheless we present evidence that Barrington's simulation technique does not carry over to the monotone world.

2. Monotone Complexity

In this section we define monotone computational models and complexity classes analogous to the usual notions of circuits, branching programs, and Turing machines. We generally will not distinguish uniform and nonuniform models unless stated explicitly. Our notation will be for a typical complexity class C, to define the monotone analogue mC. This is in contrast with the collection of functions in C that happen to be monotone, a class we refer to as $C \cap$ **mono**. We show that the standard simulations may be carried out in a monotone fashion so that most familiar class containments still hold. We consider the notion of monotone reducibility and complete problems for these classes.

2.1. Monotone Functions

Before proceeding further we ask: What is a monotone Boolean function? This is generally defined to be one where changing any input from 0 to 1 can only change the function value from 0 to 1 and not from 1 to 0. But with this definition there is no way for a monotone class to be closed under complementation, since the complements of monotone functions are not themselves monotone. This covers over an important issue since there is a natural way to redefine what we mean by monotone in such a way that mP is closed under complement yet other monotone classes remain not closed. This gives additional structural information about the monotone classes. Our definition is as follows. Call a function with the above property *positive monotone*, and the complement of a positive monotone function *negative monotone*. A function is *monotone* if it is either positive or negative monotone. We note that this usage of the terminology is more consistent with its counterpart in real analysis.

2.2. Monotone Computational Models

Monotone circuits. The standard definition of monotone circuits is to allow AND and OR gates, or perhaps some larger basis of positive monotone gates, but no negations. In order to compute negative monotone functions as well we consider a circuit to be monotone if the only negations are on the input variables, and either all of the input variables appear with negations or none of them appear with negations.

Monotone formulas. As above where all gates have outdegree 1. There has been much recent work (e.g. [KW90, RW89, RW90]) in monotone formula depth with no restriction on size, due to an exact characterization of depth by the communication complexity of certain two-party problems.

Monotone nondeterministic circuits. Nondeterministic circuits are defined to be ordinary circuits whose inputs are divided into two parts: the nondeterministic inputs and the standard inputs. A nondeterministic circuit accepts a setting of the standard inputs if there is some setting of the nondeterministic inputs which causes evaluation to 1. Each nondeterministic input bit may be input only once to the circuit; this is important in space-bounded circuits. A *monotone nondeterministic circuit* is one where the monotonicity requirement applies only to the standard inputs; that is, we may use a negation gate in the circuit as long as it depends only on nondeterministic input bits. Note that if the nondeterministic inputs were treated monotonely as well, then their only interesting setting would be all 1's or all 0's.

Monotone randomized circuits. Treat the random input bits like the nondeterministic input bits above. That is, negations are allowed which depend only on the random bits, and each random bit is input only once. In the nonuniform case, random models are generally equivalent to the deterministic model, so a proper treatment of random monotone complexity requires a uniform monotone model such as the monotone nondeterministic Turing machine model below.

Monotone nondeterministic branching programs. A nondeterministic branching program is an acyclic network of nodes and directed edges with distinguished start and finish nodes, where each edge is either labeled with the constant 1 or with a literal (a variable or a negated variable). This defines a Boolean connectivity function in a natural way: accept iff there exists a path from the start to the finish, such that all literals appearing on the path are true. *Monotone nondeterministic branching programs* are those where either none or all of the variables appear negated.

A branching program or circuit is *leveled* if the nodes are arranged in a sequence of levels with internal wires allowed only from one level to the next (inputs may be used at any level), the start node in the first level and the accept node in the last. The maximum size of a level is the *width* of the branching program or circuit.

Monotone nondeterministic Turing machines. The usual nondeterministic Turing machine (NTM) has existential nondeterminism, a read-only input tape, a read-write work tape, and a program specified by a finite set of states and transition rules. A (positive) monotone nondeterministic Turing machine (mNTM) is a NTM with the following restriction on its transition rules: whenever the machine may make a transition with a 0 on its input tape, it may make that same transition with a 1 on its input tape. Note there is no such restriction on the bits of its work tape. A negative mNTM may be defined similarly: for every 1-transition there is also a 0-transition. This mNTM model may include randomization as well, with no restriction on the use of random bits.

The above description applies to existential nondeterminism. We may generalize this to universal nondeterminism and alternation as follows. We define a monotone alternating Turing machine as the usual alternating Turing machine, but with restrictions on how it may read its input bits. An input bit x may be referenced only by either the existential construction $\exists z : (z \leq x) \wedge (\ldots)$, or by the universal construction $\forall z : (z > x) \vee (\ldots)$, where $(\ldots)$ stands for the rest of the computation. Note we do not have a uniform monotone analogue for deterministic Turing machines; however with monotone alternation we may define the classes $mALOGTIME$ and $mALOGSPACE$, which serve as uniform versions of mNC^1 and mP, respectively.

2.3. Monotone Complexity Classes

Using these models we define uniform and nonuniform monotone complexity classes.

The circuit classes mAC^i, mNC^i, and mP are standard. $mLOGCFL$ may be defined as monotone polynomial size log-depth circuits with unbounded OR gates and bounded AND gates. More generally, if the depth is $O(\lg^i n)$, we call this $mSAC^i$. mL may be defined as the class of monotone log-width polynomial size leveled circuits. From these circuit classes we may straightforwardly construct nondeterministic variants such as mNL and mNP; we note that these nondeterministic classes also have uniform definitions in terms of monotone nondeterministic Turing machines.

Polynomial size nondeterministic branching programs define the nonuniform analogue to NL; mNL is the corresponding monotone class. We remark that this is equivalent to the nondeterministic version of the circuit class mL above, where each nondeterministic bit may only be input once. Showing that the circuit model contains the branching program model is not entirely trivial; it involves representing each node internally by a binary address and the negation of that address.

Bounded width branching programs are leveled nondeterministic branching programs such that each level has size bounded by some constant. Polynomial size bounded width branching programs define the (nonuniform) class $BWBP$, and $mBWBP$ is the corresponding monotone class.

2.4.   Monotone Simulations

As mentioned before, many of the familiar simulations and containments from general complexity carry over to the monotone world. For example we may define mNC^1 either in terms of polynomial size monotone formulas or in terms of bounded fan-in log-depth monotone circuits, since it is easy to show that these are equivalent in power. Similarly mNC^1 contains $mBWBP$, since the standard simulation preserves monotonicity.

Other simulations need a little more argument. For example we have three potential models for mNL: branching program, circuit, and Turing machine. It is reassuring to argue that these models are still equivalent up to uniformity. We make these arguments in the next three paragraphs for positive monotone functions; the negative cases are similar.

Turing machine to circuit.   Given a positive monotone nondeterministic Turing machine using an $O(\lg n)$-size work tape, for a given n we want to simulate it with a monotone $O(\lg n)$-width circuit with read-once nondeterministic inputs. We follow the usual tableau construction, with the following modification: where an input bit x_i is input, we guess a nondeterministic bit z, and test if $z \leq x_i$. We then use z in the place of x_i as the tableau input. Thus the tableau may contain arbitrary negations, since they will depend only on nondeterministic bits. Also if x_i is input at many different points, we may use a different nondeterministic bit z at each point, since the monotone Turing machine is guaranteed to compute the right function even if some 1-inputs are sometimes read as 0-inputs. Finally AND together the results of all the $z \leq x_i$ comparisons with the final output of the tableau.

Circuit to branching program.   Given a leveled width-w ($w = O(\lg n)$ for mNL) monotone nondeterministic circuit with AND, OR, and 'input' gates, we wish to simulate it with a monotone polynomial size branching program. We follow a level-by-level reduction, with each vertex in the given level of the branching program corresponding to one of the 2^w possible states of the circuit level, where 'state' refers to the w-vector of truth values output by the gates in that circuit level. We design the program so that a vertex is reachable in the branching program iff the corresponding state is *less than or equal* to a state achievable in that level of the circuit. With this modification to the usual requirement, the argument proceeds straightforwardly. Logic gates

(AND and OR) and nondeterministic input gates are modeled by constant wires in the branching program. Real input gates (which in the nonmonotone case would be modeled by branching program wires labeled x_i and $\neg x_i$) may in the monotone case be modeled by edges labeled x_i and 1. We omit an argument for correctness.

Branching program to Turing machine. Essentially we need to show a monotone logspace nondeterministic Turing machine may still compute the directed st-connectivity function. Given the input as an adjacency matrix, the usual nonmonotone algorithm in fact works on a monotone Turing machine as well.

Similar simulation arguments show that our two potential models for mNP are equivalent up to uniformity. The following containments are now immediate:

Theorem 2.1 $mAC^0 \subseteq mBWBP \subseteq mNC^1 \subseteq mL \subseteq mNL \subseteq mLOGCFL \subseteq mAC^1 \subseteq mNC^i \subseteq mAC^i \subseteq mNC^{i+1} \subseteq mP \subseteq mNP.$

2.5. Trivial Monotone Classes

We say that a monotone class mC is *trivial* if it satisfies $mC = C \cap \mathbf{mono}$.

Given any kind of (existential) nondeterministic machine, the corresponding positive monotone machine is one where for each 0-transition—a transition allowed when some input bit $x_i = 0$—the corresponding 1-transition is also allowed. This generalizes our definition for mNTMs, similar definitions apply for negative monotone machines. We say a nondeterministic machine is read-once if it reads each input bit at most once on any computation path; the bits do not have to be read in the same order on each path.

Theorem 2.2 *Given complexity class* C *defined in terms of a read-once nondeterministic machine, the corresponding monotone class* mC *is trivial.*

Proof : Given a (nonmonotone) nondeterministic machine accepting a positive monotone language L (i.e. given $x \in L$, if we replace any 0 in x by a 1 then the resulting string is still in L), we need to show L is also accepted by a monotone machine. We convert the given automaton into a monotone automaton by simply allowing a 1-transition whenever the original machine allowed a 0-transition.

Suppose the new monotone machine accepts string y. Then the computation path accepting y may have used some of the new 1-transitions, but if we replace each such 1 in y by a 0, we get a string x that was accepted by the original automaton, hence $x \in L$. Since L is monotone and $x \in L$, we have $y \in L$; thus the monotone machine accepts exactly L.   □

It follows that $mREG$ (via nondeterministic finite state automata), $mCFL$ (via nondeterministic push-down automata), and mNP (since an mNP machine may copy its entire input to its internal work tape) are all trivial classes.

We note that a similar argument applies in some circuit models, such as the circuit version of mNP or monotone read-once branching programs.   Finally we note the triviality of $AC[2]$ circuits (polynomial size CNF or DNF formulas).

Theorem 2.3 mAC[2] *is trivial.*

Proof :   For an AND of ORs computing a positive monotone function, it suffices to replace each negated input by a 0. Similarly for an OR of ANDs, it suffices to replace each negated input by a 1.   □

This method fails for depth three; we do not even know if $AC[3] \cap$ **mono** $\subset$ mP.

2.6.   Rephrasing Known Results

We may rephrase some known results on monotone complexity as follows:

Theorem 2.4 ([Raz85a]) $mP \neq mNP$.

Theorem 2.5 ([Raz85b]) $mP \neq P \cap$ **mono**.

Theorem 2.6 ([AG87]) $mAC^0 \neq AC^0 \cap$ **mono**.

Theorem 2.7 ([KW90]) $mNC^1 \neq mNL$.

Theorem 2.8 ([Yao89]) $mTC^0 \neq mNC^1$.

Theorem 2.9 ([RW90]) $mNC^1 \cap$ **mono** $\not\subset$ mNC.

2.7.   Monotone Separations

As listed above, there are many separations of monotone classes with no corresponding separation of the corresponding nonmonotone classes; so far no monotone separation methods have successfully carried over. It seems then that nonmonotone separations are at least as hard as monotone separations. On the other hand showing a containment between classes is a stronger result in the monotone world than in general, since we need to check that monotonicity is preserved. Here we argue that these intuitions are correct, at least for a wide range of classes.

For reductions we use polynomially bounded Boolean projections, from Skyum and Valiant [SV85]. Given two Boolean function families $\{f_n\}$ and $\{g_n\}$, we say that f is a projection of g if for each n there exists $p(n)$ and σ such that $f_n(x) = g_{p(n)}(\sigma(x))$. Here $p(n)$ is bounded by some polynomial and $\sigma(x)$ is a substitution; i.e. every argument of g is either a constant, an argument of f, or a negated argument of f. If we disallow negative (or positive) argument substitutions, then we say that f is a monotone projection of g.

If C is a complexity class and mC is its monotone analogue, then we are interested in the following question: does there exist a monotone function f complete for C under general projections and also complete for mC under monotone projections? We say such an f is a *monotone complete function* for C. This notion is motivated by the following observation.

Theorem 2.10 *If* C_1 *and* C_2 *are complexity classes with monotone complete functions, then* $mC_1 \subseteq mC_2$ *implies* $C_1 \subseteq C_2$.

Proof :   Let f_1 and f_2 be the corresponding monotone complete functions. Since f_1 is a monotone projection of f_2, it is also a general projection.   □

For all classes we have considered, there is such a function. For circuit classes the canonical monotone complete function is the circuit value problem for an appropriate monotone universal circuit. For example a monotone complete problem for NC^1 is the balanced formula of $O(\lg n)$ depth with alternating levels of ANDs and ORs (with distinct variables at the inputs) since this may be restricted to compute any log-depth formula. Simple universal circuit constructions (simulating level-by-level) apply to AC^i, SAC^i, mL, mP, and mNP, while mNC^i may be handled by simulating $O(\lg n)$ levels at a time using the mNC^1 universal formula as a building block.

For nondeterministic branching program classes the appropriate graph connectivity problem (where each directed edge is represented by a variable) is monotone complete. Directed st-connectivity is monotone complete for mNL. A similar statement (for each bounded width) holds for bounded width digraph connectivity and $mBWBP$.

3.   Separating mNL from co-mNL

3.1.   Semi-Unbounded Circuits

We first consider the semi-unbounded circuits of Venkateswaran [Ven87]. As above, let SAC^k denote the class of polynomial-size $O(\lg^k n)$-depth circuits with bounded fan-in AND-gates, unbounded fan-in OR-gates, and negations allowed on the inputs (we borrow this notation from [BCD+89], see [Bus87] and [Ven87] for properties of $SAC^1 = LOGCFL$). Without loss of generality, the bounded fan-in gates have fan-in two. Similarly define co-SAC^k with bounded OR-gates and unbounded AND-gates. Define the monotone variants $mSAC^k$ and co-$mSAC^k$ similarly but with monotone inputs. Computing the directed connectivity function by repeated squaring of the adjacency matrix, it follows that $NL \subseteq SAC^1$ and $mNL \subseteq mSAC^1$.

A result of Borodin et. al. [BCD+89], closely related to Immerman's result, states that $SAC^k = co$-SAC^k for $k \geq 1$. We show that $mSAC^1 \neq co$-$mSAC^1$ in the monotone model.

3.2.   The Separation

Let **ustconn** denote the undirected st-connectivity function. For a graph of n vertices with two distinguished vertices s and t, **ustconn** has an input variable for each of the $\binom{n}{2}$ edges; clearly **ustconn** $\in mNL$. Karchmer and Wigderson [KW90] showed that any bounded fan-in monotone formula computing **ustconn** requires depth $\Omega(\lg^2 n)$. We extend their proof to show that any co-$mSAC$ circuit for **ustconn** also requires depth $\Omega(\lg^2 n)$. It then follows that **ustconn** $\notin co$-mNL, i.e. large fan-in AND-gates do not help to compute the **ustconn** function.

An l-$path$ in the graph is a sequence of l vertices (perhaps with repetitions) defining a path from s to t. There are n^l such paths. A $cut\ graph$ is formed by dividing the n vertices into two sets, with s and t in different sets, and putting edges between vertices in the same set. There are 2^{n-2} such cuts. Say that a positive monotone function is an (n, l, α, β) approximator if it outputs 1 for at least fraction α of l-paths, and it outputs 0 for at least fraction β of cut-graphs.

Consider a positive *co-mSAC* circuit C which is an (n, l, α, β) approximator. If the top gate of C is an OR-gate, then one of the two children subcircuits is an $(n, l, \alpha/2, \beta)$ approximator. If the top gate of C is an AND-gate with fan-in at most $d(n)$, then one of the children subcircuits is an $(n, l, \alpha, \beta/d(n))$ approximator. Iterating this process down the circuit for k levels leads to an $(n, l, \alpha/2^k, \beta/d(n)^k)$ approximator.

To show that $\Omega(\lg^2 n)$ depth is necessary, periodically restrict the circuit to get a better approximator to a slightly smaller instance of **ustconn**. This is done with a random restriction $\rho \in \mathcal{R}_k$, which chooses k vertices other than s or t, splits these k into two sets, and then contracts one set with s and the other set with t. We use the following lemma of Boppana [BS90]:

Lemma 3.1 *Let positive monotone function f be an (n, l, α, β) approximator. Suppose $100l/\alpha \leq k \leq n/(100l)$ and $\beta \geq 2^{-n/(100k)}$. Then there is a restriction $\rho \in \mathcal{R}_k$ such that f_ρ is an $(n - k, l/2, \sqrt{\alpha}/2, \beta k/(2n))$ approximator.*

Note α increases greatly while n, l, and β decrease in a controlled way. If f is computed by a positive monotone circuit, then f_ρ is computed by the same circuit with some trivial monotone modification of the inputs. We now prove our theorem:

Theorem 3.2 *Let C be a positive monotone circuit computing **ustconn** with fan-in two OR-gates and fan-in $d(n)$ AND-gates, where $d(n) = 2^{\sqrt{n}/(\lg^2 n) - 15}$. Then C has depth greater than $(\lg^2 n)/100$.*

Proof : Suppose C has depth $(\lg^2 n)/100$. Divide C into $(\lg n)/10$ blocks of $(\lg n)/10$ levels each. Let $l = n^{1/4}$ and let $k = \sqrt{n}$. Note C is an $(n, l, 1, 1)$ approximator.

Repeatedly explore down one block of C, and then apply the lemma to find a restricted subcircuit with boosted α. At no point during the exploration does α fall below $(n^{-1/5})/4$. When we reach the bottom, we have found a monotone depth-0 circuit that is an $(n', l', \alpha', \beta')$ approximator, where $n' = n - \sqrt{n}(\lg n)/10$, $l' = l/n^{1/10} = n^{1/5}$, $\alpha' = (n^{-1/10})/4$, and $\beta' = d(n)^{-(\lg^2 n)/100}(2\sqrt{n})^{-\lg n/10}$.

A depth-0 circuit is either a constant or a single positive variable. Since $\alpha' > 0$ and $\beta' > 0$, the gate is not a constant. But a single variable cannot accept such a large fraction α' of l-paths. Hence we have a contradiction.

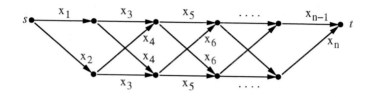

Figure 1. A width-2 $mBWBP$ function outside mAC^0.

The only new thing to check in this proof is that $\beta' \geq 2^{-n/(100k)}$, which is satisfied by our choice of $d(n)$.   □

Corollary 3.3 co-mSAC *circuits for* **ustconn** *require depth* $\Omega(\lg^2 n)$.

Corollary 3.4 co-mNL *branching programs for* **ustconn** *have size* $n^{\Omega(\lg n)}$.

Straightforward constructions show both the above lower bounds are tight. Suitably adjusting the constants in the above proof shows that even if the AND-gates have fan-in $2^{n^{1-\epsilon}}$, the circuit must still have depth $\Omega(\epsilon^2 \lg^2 n)$. Note the depth lower bound does not depend on circuit size, except as an upper bound on $d(n)$. Also since **ustconn** is in RL, we have (nonuniformly)

Corollary 3.5 $L \cap$ **mono** $\not\subseteq$ mNL.

4.   Towards Separating $mBWBP$ from mNC^1

Motivated by Barrington's surprising construction showing $BWBP = NC^1$, a natural question is whether $mBWBP$ (monotone bounded-width branching programs) equals mNC^1.

We remark that $mBWBP$ is equivalent to monotone straight-line Boolean programs with a bounded number of registers (if the program may take the AND of two registers, then the branching program width is exponential in the number of registers). Also it is clear that $mBWBP$ is strictly greater than mAC^0, since a width-2 branching program (see figure 1) computes a function which, if it were in mAC^0, would put parity in AC^0.

To show $mBWBP$ is strictly contained in mNC^1, one could consider a complete formula for mNC^1, or try to show that $mBWBP$ is not closed under complementation. We conjecture that the majority function (known to be in mNC^1 [Val84, AKS83]) is not in $mBWBP$. Proving this would require some new technique, since previous methods in monotone complexity do not seem to apply to slice functions. We present two preliminary results.

4.1. A Lower Bound on Size

The monotone branching program model is very close to the (undirected) notion of relay networks studied by Shannon and Moore [MS56] in the context of probabilistic amplification, and identical to "relay-diode bipoles" as studied by Markov [Mar62].

Both papers give versions of the following simple lower bound. Given a positive monotone function f, let the *length* of f be the minimum size of any minterm, and let the *width* of f be the minimum size of any maxterm. Denote these quantities by $l(f)$ and $w(f)$.

Theorem 4.1 (Markov) *Any monotone nondeterministic branching program computing f has at least $l(f) \cdot w(f)$ variable edges.*

Proof :   Given such a branching program, for each node u define $d(u)$ as the minimum number of variables that need to be set to 1 to establish a directed path from the start node s to u. In particular $d(t) = l(f)$ for the finish node t.

For $0 \leq i \leq l(f)$ let S_i be the set of nodes u such that $d(u) = i$. If u is connected to v by a constant 1 edge (i.e. not a variable edge) then $d(u) \geq d(v)$, hence there are no constant edges from S_i to S_j for $i < j$. Thus for $0 \leq i < l(f)$, the set of variable edges out of S_i forms an (s,t)-cut of the branching program. Such a cut contains a maxterm, hence at least $w(f)$ distinct variable edges.   □

Consider the threshold-k function $T_{n,k}$, which is 1 iff at least k of its n inputs are 1. Then $l(T_{n,k}) = k$ and $w(T_{n,k}) = n-k+1$, so every monotone branching program has at least $k \cdot (n - k + 1)$ variable edges. This bound is tight for unbounded width branching programs, for example see figure 2.

Corollary 4.2 *Monotone bounded width branching programs for majority have length $\Omega(n^2)$.*

We know of no better lower bound for the bounded width case. By contrast, the best lower bounds on general bounded width branching programs for explicit functions are superlinear by only logarithmic factors, using much more involved techniques.

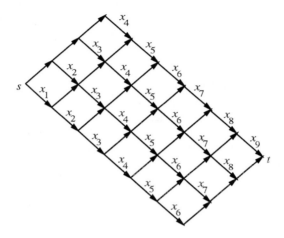

Figure 2. The naive threshold-k branching program has $k \cdot (n - k + 1)$ variable edges (here $n = 9$, $k = 6$).

4.2. There is no Monotone Barrington Gadget

Here we show that there is no monotone gadget (of any length or width) like Barrington's gadget composed of four 5-cycles [Bar89]. We need a few definitions to state this precisely.

For constant width w, define a (positive) monotone width-w branching program as in section 2.2.. The branching program computes a 1 iff there exists a directed path from the start node s to the finish node t such that every variable on the path is set to 1.

For levels i and j, $i \leq j$, let $C(i, j)$ be the w by w matrix of Boolean functions describing the connectivity from level i to j, i.e. $C(i, j)_{u,v}(x)$ is true iff there exists a path (valid with respect to x) from node u in level i to node v in level j. Given the simple single-step matrices $C(i) = C(i, i+1)$, all the others are defined by Boolean matrix multiplication: $C(i, j) \cdot C(j, k) = C(i, k)$.

We consider a constant Boolean matrix M (all 0's and 1's) to be a monotone function from P to P, where P is the poset of all 2^w subsets of the w nodes in a level. Given a subset I of $\{1, \ldots, w\}$, M maps I to $M(I)$, the set of all nodes reachable from I via M. More precisely, if χ_I is the characteristic row vector for I, then the characteristic vector for $M(I)$ is $\chi_I \cdot M$. Note that by this definition the applicative order of matrices is from left to right, so that $(AB)(I) = B(A(I))$.) The argument that follows will apply in the more general setting where P is some finite poset and the M's are monotone functions from P to P.

Given a pair of constant matrices (M_0, M_1) and a Boolean variable z, let $M(z)$ denote the variable matrix M_z. If $M_0 = M_1$, we say that $M(z)$ is trivial. If $M_0 \leq M_1$ (i.e. $M_0(I) \leq M_1(I)$ for all $I \in P$) we say that $M(z)$ is monotone.

Barrington's construction [Bar89] (and similar later constructions) using nonsolvable groups may be stated as finding a finite poset P, nontrivial (and nonmonotone) $M(z)$, and constant matrices A_i, B_i ($0 \leq i \leq 4$) such that the following hold:

$$M(x \wedge y) = A_0 M(x) A_1 M(y) A_2 M(x) A_3 M(y) A_4,$$
$$M(x \vee y) = B_0 M(x) B_1 M(y) B_2 M(x) B_3 M(y) B_4.$$

Given such equations, it is straightforward to show how $BWBP$ may simulate NC^1 with at most quadratic blowup from formula size to the length of the program. Similarly if one can find more equations for a larger basis of functions (e.g. an equation for $M(x \wedge (y \vee z))$) then it is possible to reduce the quadratic blowup to some smaller exponent.

Our result here is that if $M(z)$ is restricted to be monotone, then there is no such pair of equations of any length. For any finite poset P, any $k, l \geq 2$, any sequences of constant matrices A_i, B_j ($0 \leq i \leq k$, $0 \leq j \leq l$) and variables $v_i, w_j \in \{x, y\}$ ($1 \leq i \leq k$, $1 \leq j \leq l$), there is no nontrivial monotone solution $M(z)$ to the equations

$$M(x \wedge y) = A_0 M(v_1) A_1 M(v_2) A_2 \cdots M(v_k) A_k, \tag{1}$$
$$M(x \vee y) = B_0 M(w_1) B_1 M(w_2) B_2 \cdots M(w_l) B_l. \tag{2}$$

Define the *rank* of a constant matrix M to be the size of the range of M as a function on P: $r(M) = |M(P)|$ where $M(P) = \{M(I) : I \in P\}$. For a two-valued matrix $M(z)$, define $r_0(M(z)) = r(M_0)$, $r_1(M(z)) = r(M_1)$, and $r(M(z)) = |M_0(P) \cup M_1(P)|$.

Theorem 4.3 *If $M(z)$ is a monotone solution to equations 1 and 2, then $M(z)$ is trivial.*

Proof : Without loss of generality the last variable in each equation is y, i.e. $v_k = w_l = y$. Setting $y = 1$ in the first (AND) equation yields $M(x) = F(x) M_1 A_k$, where $F(x)$ is some arbitrary monotone matrix function of x. Thus for all inputs I to $M(x)$ and no matter how we vary x, the range of

$M(x)$ is no larger than the range of M_1, i.e. $r(M(z)) \leq r_1(M(z))$. By a similar argument applied to the second (OR) equation with $y = 0$, we have $r(M(z)) \leq r_0(M(z))$. But since $r(M(z))$ is at least as large as both $r_0(M(z))$ and $r_1(M(z))$, they all represent the same range R of size r, regardless of whether z is 0, 1, or variable.

We consider any $M(v_i)$ in the first equation for $i \geq 2$. Since the whole equation has range r for any setting of x and y, and since the number of different inputs that $M(v_i)$ can receive from $M(v_{i-1})A_i$ is also at most r, $M(v_i)$ must be a one-to-one function from its r possible inputs to r possible outputs R.

More precisely, let D_i be the domain (set of all possible inputs) of $M(v_i)$. We argued that D_i does not depend on the settings of x and y. Now we consider M_0 and M_1 restricted to D_i.

Lemma 4.4 *Given bijections N_0 and N_1 from a set D to finite poset R, such that $N_0(x) \leq N_1(x)$ for all $x \in D$, then $N_0 = N_1$.*

Proof :   Necessarily $|D| = |R|$. Define $f = N_1 N_0^{-1}$ mapping R to R, then $y \leq f(y)$ for all $y \in R$. If for some y we have $y < f(y)$, then it follows that $f(y) < f(f(y))$. Iterating yields an infinite chain in R, violating the finiteness of R. Hence f is the identity and $N_0 = N_1$.   □

Applying the lemma to $D = D_i$, with N_0 and N_1 the restrictions of M_0 and M_1 to D, we see that $M_0 = M_1$ on D_i, i.e. $M(v_i)$ doesn't really depend on v_i at all!

Since this applies to all $i \geq 2$, equation 1 reduces to $M(x \wedge y) = A_0 M(v_1) C$ where C is a constant matrix. By setting v_1 (say $v_1 = y$) to 1, we get $M(x) = A_0 M_1 C$, so $M(x)$ is trivial.   □

The above argument generalizes to "grammars" of matrices: if we have a finite collection of two-valued matrices $M^1(z)$, $M^2(z), \ldots$, such that each $M^k(x \wedge y)$ and $M^k(x \vee y)$ can be expanded as an equation in the others, then they are all trivial.

Of course a separation of $mBWBP$ from mNC^1 would imply the above theorem, but perhaps this simple rank method (or the metric method of Theorem 4.1) will help resolve the general question.

We remark that in the special case where the underlying poset P is a chain, then the space of all monotone functions on P form an aperiodic monoid, and programs over such a monoid are computable in AC^0 by the work of

Barrington and Thérien [BT88]. Similarly if we restrict to a solvable monoid of monotone functions on P, we get a monotone analogue of ACC^0, which may be more tractable than $mBWBP$.

5.   Conclusion

We conclude with some open problems in monotone complexity.

1. Find a straightforward uniform monotone analogue for deterministic Turing machines. For example we still have no uniform model for mL.

2. Suggested by Larry Stockmeyer: show there is no polynomial size monotone projection from the directed st-connectivity function to the undirected st-connectivity function.

3. Many of the inclusions in Theorem 2.1 are still not known to be proper (e.g. can we separate higher levels of the mNC hierarchy or separate mNC from mP). Recently [GS91, Gri91] we have succeeded in separating mL from mNC^1, again by the communication complexity method.

4. Is $AC^0 \cap \mathbf{mono} \subseteq mP$?   That is, is there any nontrivial monotone upper bound on the complexity of monotone functions which have very low nonmonotone complexity? Since $NC^2 \cap \mathbf{mono} \not\subseteq mP$ (by the matching function [Raz85a]), we cannot hope for much more in mP. The only limit we know here is that $mNP = NP \cap \mathbf{mono}$.

5. Separate mL from $mNL \cap co\text{-}mNL$. A candidate function is directed planar st-connectivity where s and t are on the outer face; it suffices to consider grid graphs. An initial question is whether this function requires $\Omega(\lg^2 n)$ depth.

References

[AG87]   Ajtai M., Gurevich Y., *Monotone versus positive. J. ACM, Vol. 34 (1987), pp. 1004–1015.*

[AKS83]   Ajtai M., Komlos J., Szemeredi E., *An $O(n \log n)$ Sorting Network. Combinatorica, Vol. 3 (1983), pp. 1–19.*

[AB87]   Alon N., Boppana R., *The monotone circuit complexity of boolean functions. Combinatorica, Vol. 7 (1987), pp. 1–22.*

[And85]   Andreev A., *On a method for obtaining lower bounds for the complexity of individual monotone functions. Doklady Akademii Nauk SSSR, Vol. 282 (1985), pp. 1033–1037. English translation in Soviet Mathematics Doklady, Vol. 31 (1985), pp. 530–534.*

[BFS86]   Babai L., Frankl P., Simon J., *Complexity classes in communica-tion complexity theory. Proceedings of 27th Annual IEEE Symposium on Foundations of Computer Science, (1986), pp. 337–347.*

[Bar89]   Barrington D.A., *Bounded-width polynomial-size branching programs recognize exactly those languages in* NC^1. *J. Comput. System Sci., Vol. 38 (1989), pp. 150–164.*

[BT88]   Barrington D.A., Thérien D., *Finite monoids and the fine structure of* NC^1. *J. ACM, Vol. 35 (1988), pp. 941–952.*

[BS90]   Boppana R., Sipser M., *The complexity of finite functions. In Handbook of Theoretical Computer Science, Vol. A, Elsevier and MIT Press, 1990, pp. 757–804.*

[BCD$^+$89] Borodin A., Cook S., Dymond P., Ruzzo W., Tompa M., *Two applications of inductive counting for complementation problems. SIAM J. Computing, Vol. 18 (1989), pp. 559–578.*

[Bus87]   Buss S., *The Boolean formula value problem is in ALOGTIME. Proceedings of 19th Annual ACM Symposium on Theory of Computing, 1987, pp. 123–131.*

[FSS84]   Furst M., Saxe J., Sipser M., *Parity, circuits, and the polynomial time hierarchy. Mathematical Systems Theory, Vol. 17 (1984), pp. 13–27.*

[Gri91]   Grigni M., *Structure in Monotone Complexity. Technical Report MIT/LCS/TR-520, Massachusetts Institute of Technology, 1991.*

[GS91]   Grigni M., Sipser M., *Monotone separation of logspace from* NC^1. *Proceedings of the Sixth Annual IEEE Conference on Structure in Complexity Theory, 1991, pp. 294–298.*

[Imm88]   Immerman N., *Nondeterministic space is closed under complementation. SIAM J. on Computing, Vol. 17 (1988), pp. 935–938.*

[KW90]   Karchmer M., Wigderson A., *Monotone circuits for connectivity require super-logarithmic depth. SIAM J. Discrete Mathematics, Vol. 3 (1990), pp. 255–265.*

[Mar62]   Markov A.A., *Minimal relay-diode bipoles for monotonic symmetric functions. Problemy Kibernetiki, Vol. 8 (1962), pp. 117–121. English translation in Problems of Cybernetics, Vol. 8 (1964), pp. 205–212.*

[MS56]   Moore E.F., Shannon C.E., *Reliable circuits using less reliable re-lays*. J. of the Franklin Inst., Vol. 262 (1956), pp. 191–208 and 281–297.

[RW89]   Raz R., Wigderson A., *Probabilistic communication complexity of boolean relations*. Proceedings of 30th Annual IEEE Symposium on Foundations of Computer Science, 1989, pp. 562–573.

[RW90]   Raz R., Wigderson A., *Monotone circuits for matching require linear depth*. Proceedings of the 22nd Annual ACM Symposium on Theory of Computing, 1990, pp. 287–292.

[Raz85a] Razborov A.A., *Lower bounds on the monotone complexity of some Boolean functions*. Doklady Akademii Nauk SSSR, Vol. 281 (1985), pp. 798–801. English translation in Soviet Mathematics Doklady, Vol. 31 (1985), pp. 354–357.

[Raz85b] Razborov A.A., *A lower bound on the monotone network complexity of the logical permanent*. Matematicheskie Zametki, Vol. 37 (1985), pp. 887–900. English translation in Mathematical Notes of the Academy of Sciences of the USSR, Vol. 37 (1985), pp. 485–493.

[SV85]   Skyum S., Valiant L., *A complexity theory based on boolean algebra*. J. ACM, Vol. 32 (1985), pp. 484–502.

[Sze87] Szelepcsényi R., *The method of forcing for nondeterministic automata*. Bull. European Ass. for Theoretical Computer Science, Vol. 33 (1987), pp. 96–100.

[Tar88] Tardos E., *The gap between monotone and nonmonotone circuit complexity is exponential*. Combinatorica, Vol. 8 (1988), pp. 141–142.

[Val84] Valiant L., *Short monotone formulae for the majority function*. Journal of Algorithms, Vol. 5 (1984), pp. 363–366.

[Ven87] Venkateswaran H., *Properties that Characterize LOGCFL*. Proceedings of 19th Annual ACM Symposium on Theory of Computing, 1987, pp. 141–150.

[Yao89] Yao A., *Circuits and local computation*. Proceedings of 21st Annual ACM Symposium on Theory of Computing, 1989, pp. 186–196.

On Submodular Complexity Measures

A. A. Razborov *

1. Introduction

In recent years several methods have been developed for obtaining superpolynomial lower bounds on the monotone formula and circuit size of explicitly given Boolean functions. Among these are the method of approximations [3, 4, 1, 7, 15, 2], the combinatorial analysis of a communication problem related to monotone depth [9, 12] and the use of matrices with very particular rank properties [13]. Now it can be said almost surely that each of these methods would need considerable strengthening to yield nontrivial lower bounds for the size of circuits or formulae over a complete basis. So, it seems interesting to try to understand from the formal point of view what kind of machinery we lack.

The first step in that direction was undertaken by the author in [14]. In that paper two possible formalizations of the method of approximations were considered. The restrictive version forbids the method to use extra variables. This version was proven to be practically useless for circuits over a complete basis. If extra variables are allowed (the second formalization) then the method becomes universal, i.e. for *any* Boolean function f there exists an approximating model giving a lower bound for the circuit size of f which is tight up to a polynomial. Then the burden of proving lower bounds for the circuit size shifts to estimating from below the minimal number of covering sets in a particular instance of *"MINIMUM COVER"*. One application of an analogous model appears in [5] where the first nonlinear lower bound was proven for the complexity of *MAJORITY* with respect to switching-and-rectifiers networks.

R. Raz and A. Wigderson in [11, 12] gave an indication that the communication problem of Karchmer and Wigderson [9] over the standard basis with negations also behaves very differently from its monotone analogue. Namely,

---

*Steklov Mathematical Institute, 117966, GSP-1, Vavilova, 42, Moscow, USSR

they showed that the probabilistic complexity of this problem for *any* specific Boolean function is $O(\log n)$ whereas the probabilistic complexity of the problem related to the *monotone* depth of *"PERFECT MATCHING"* is $\Omega(n)$.

In the present paper we study in this fashion the third method among those listed in the first paragraph, i.e. the method which relies upon constructing a matrix whose rank is much bigger than the ranks of certain submatrices of this matrix. We show that this method cannot even give nonlinear lower bounds over the standard basis with negations. This answers an open question from [13]. On the other hand we observe that if the matrix is allowed to be partial, then the method becomes very powerful.

Actually, we can treat a natural class of methods which contains the one from [13]. To say exactly what this class is, we recall (see e.g. [16, §8.8]) the notion of a formal complexity measure. Namely, a nonnegative real-valued function μ defined on the set of all Boolean functions in n variables is a *formal complexity measure*[1] if

$$\mu(x_i) \leq 1, \quad \mu(\neg x_i) \leq 1 \quad (1 \leq i \leq n); \tag{1}$$

$$\mu(f \vee g) \leq \mu(f) + \mu(g) \quad \text{for each } f, g; \tag{2}$$

and

$$\mu(f \wedge g) \leq \mu(f) + \mu(g) \quad \text{for each } f, g. \tag{3}$$

Restricting the domain of μ and arguments in (1-3) to the set of monotone functions, we obtain the definition of a *formal complexity measure on monotone functions*. Obvious induction shows that for any formal complexity measure μ we have $\mu(f) \leq L(f)$ ($L(f)$ is the formula size of f) and similarly for the monotone case. Actually, proofs of many known lower bounds on $L(f)$ can be viewed as inventing clever formal complexity measures which can be nontrivially bounded from below at some explicitly given Boolean functions (see e.g. the re-formulation of the Khrapchenko bound [6] given by M. Paterson [16, §8.8]).

We will see that the matrix method from [13] can also be easily reformulated in terms of formal complexity measures. Moreover, it turns out (Theorem 1 below) that the resulting measures μ satisfy the *submodularity condition*

$$\mu(f \wedge g) + \mu(f \vee g) \leq \mu(f) + \mu(g) \quad \text{for each } f, g \tag{4}$$

which is stronger than both (2) and (3). We call a formal complexity measure μ *submodular* if (4) holds and similarly for the monotone case. The results

---

[1]In this definition we have removed several unnecessary conditions from [16].

from [13] imply the existence of submodular formal complexity measures on monotone functions which take on values of size $n^{\Omega(\log n)}$.

The main result of this paper (Theorem 2) says that *all values of any submodular formal complexity measure* (on the set of *all* Boolean functions in n variables) *are bounded from above by* $O(n)$.

It is worth noting that the proof of Theorem 2 makes use of the same random circuit $\mathbf{C}$ which was previously used in the proof of Lemma 3.1 from [14] for breaking down the restrictive version of the method of approximations. It seems that this circuit can act as a hard test for different ideas aimed at proving lower bounds on the size of circuits or formulas over a complete basis.

2.   Definitions and example of submodular complexity measures

Throughout the paper B^n denotes an n-dimensional Boolean cube and F_n [F_n^{mon}] the set of all Boolean functions [the set of all monotone Boolean functions respectively] in n variables. For $u \in B^n$, $1 \leq i \leq n$, u^i means the ith bit in u. Let $X_i^\epsilon \rightleftharpoons \{u \in B^n | u^i = \epsilon\}$ for $1 \leq i \leq n$, $\epsilon \in \{0,1\}$. Given a variable x_i, set $x_i^1 \rightleftharpoons x_i$; $x_i^0 \rightleftharpoons (\neg x_i)$. Given $f \in F_n$, $U \subseteq B^n$, $\epsilon \in \{0,1\}$, the statement $\forall u \in U \ (f(u) = \epsilon)$ will be written in the simplified form $f(U) = \epsilon$. By a *formula* (over the standard basis) we mean an expression of the propositional calculus constructed (following the usual rules and conventions) from variables $x_1, x_2, ..., x_n$ with connectives $\vee$, $\wedge$, $\neg$; every formula $\phi(x_1, x_2, ..., x_n)$ computes in a natural way some function from F_n. The *size* $s(\phi)$ of a formula ϕ is the total number of occurrences of variables in ϕ. Using de Morgan's laws we can transform every formula into a *formula with tight negations* (i.e., a formula in which negations occur only in the form $(\neg x_i)$) without increasing its size. Given $f \in F_n$, the *formula size* $L(f)$ is min $\{s(\phi)|\phi$ computes $f\}$. A formula is *monotone* if it contains no negations at all; the *monotone formula size* $L_{\mathrm{mon}}(f)$ of an $f \in F_n^{\mathrm{mon}}$ is defined by analogy with $L(f)$.

We say that a function $\mu : F_n \longrightarrow \mathbb{R}^+$ is a *submodular formal complexity measure* (or *submodular complexity measure* for short) if it satisfies (1) and (4) (and hence also satisfies (2) and (3)). A function $\mu : F_n^{\mathrm{mon}} \longrightarrow \mathbb{R}^+$ is a *submodular complexity measure on* F_n^{mon} if it satisfies the first condition in (1) and satisfies (4) whenever $f, g \in F_n^{\mathrm{mon}}$. In the rest of the section we consider an example of submodular complexity measures.

Let $U, V \subseteq B^n$, $U \cap V = \varnothing$. A *rectangle* (over U, V) is an arbitrary subset of the Cartesian product $U \times V$ which has the form $U_0 \times V_0$ where $U_0 \subseteq U$,

$V_0 \subseteq V$. Every set $\mathcal{R}$ of rectangles such that $\cup \mathcal{R} = U \times V$ will be called a *covering* (over U, V). The *canonical* covering $\mathcal{R}_{\text{can}}(U, V)$ is defined as follows:

$$\mathcal{R}_{\text{can}}(U, V) \rightleftharpoons \{R_{01}, R_{02}, ..., R_{0n}, R_{11}, R_{12}, ..., R_{1n}\}$$

where $R_{\epsilon i} \rightleftharpoons (U \cap X_i^{\epsilon}) \times (V \cap X_i^{1-\epsilon})$ $(1 \leq i \leq n, \epsilon \in \{0, 1\})$.

By a *matrix over U, V* we mean a matrix over a field k whose rows are indexed by elements of the set U and columns by elements of the set V. Given a rectangle R, we denote by A_R the corresponding submatrix of a matrix A. The following result was proved in [13].

Proposition 1 [13] *For any $U, V \subseteq B^n$ and $f \in F_n$ such that $f(U) = 0$, $f(V) = 1$ and any non-zero matrix A over U, V (over an arbitrary field k), the inequality*

$$L(f) \geq \frac{\text{rk}(A)}{\max_{R \in \mathcal{R}_{\text{can}}(U,V)} \text{rk}(A_R)} \tag{5}$$

holds.

Let us understand that this lower bound is essentially the bound provided by a submodular complexity measure μ. For arbitrary $f \in F_n$, define the rectangle R_f over U, V by

$$R_f \rightleftharpoons (U \cap f^{-1}(0)) \times (V \cap f^{-1}(1)).$$

Let

$$\mu(f) \rightleftharpoons \frac{\text{rk}(A_{R_f})}{\max_{R \in \mathcal{R}_{\text{can}}(U,V)} \text{rk}(A_R)}.$$

Theorem 1 *a)* μ *is a submodular complexity measure;*
b) if $f(U) = 0$, $f(V) = 1$ then $\mu(f)$ equals the right-hand side of (5).

Proof :   a) We have to check (1) and (4). (1) trivially follows from the definitions because if $f = x_i^{\epsilon}$ then $R_f = R_{i,1-\epsilon} \in \mathcal{R}_{\text{can}}(U, V)$. For proving (4) consider the linear space k^U with the set U embedded into it as the basis. Using the matrix A we can also map V to k^U ($v \in V$ goes to the corresponding column of A). For $W \subseteq U \cup V$, denote by $\rho(W)$ the dimension of the subspace generated in k^U by the image of W via the mapping described. Then ρ is the rank function of a (linear) matroid on $U \cup V$ and hence is submodular. Now, $\text{rk}(A_{R_f})$ can be expressed in the form

$$\text{rk}(A_{R_f}) = \rho((U \cup V) \cap f^{-1}(1)) - |U \cap f^{-1}(1)|.$$

The submodularity of $\mathrm{rk}(A_{R_f})$ (and hence $\mu(f)$) follows from the submodularity of ρ.

b) is trivial.   □

Similar results hold for the monotone case if we place on $U, V \subseteq B^n$ the restriction

$$\forall u \in U \,\forall v \in V \,\exists i \,(u^i = 0 \,\&\, v^i = 1)$$

(which is stronger than just $U \cap V = \varnothing$) and replace $\mathcal{R}_{\mathrm{can}}(U, V)$ by

$$\mathcal{R}_{\mathrm{mon}}(U, V) \rightleftharpoons \{R_{01}, R_{02}, ..., R_{0n}\}.$$

It was shown in [13] that any 0-1 matrix A for which the rank lower bound of Mehlhorn and Schmidt [10] gives a *superlinear* gap between $\mathrm{DCC}(A)$ and $\max\,(\mathrm{NCC}(A), \mathrm{NCC}(\neg A))$ ($\mathrm{DCC}(A)$ and $\mathrm{NCC}(A)$ are deterministic and non-deterministic communication complexities of A respectively), can be used to construct a monotone Boolean function for which the monotone analogue of Proposition 1 gives a *superpolynomial* lower bound on its monotone formula size. In particular, the matrices presented in [10] lead to the bound $n^{\Omega(\log n/\log \log n)}$ and the matrices from [8] and [13] lead to the bound $n^{\Omega(\log n)}$. Applying the monotone analogue of Theorem 1, we obtain

Corollary 1 *There exist submodular complexity measures on F_n^{mon} which take on values of size at least $n^{\Omega(\log n)}$.*

3.   Main result

The reader is invited to compare the following theorem (which is the main result of this paper) with Corollary 1 above and the proof of this theorem with the proof of Lemma 3.1 in [14].

Theorem 2 *For each submodular complexity measure μ (on F_n) and each $f_n \in F_n$ we have $\mu(f_n) \leq O(n)$.*

Proof :   Let $\mathbf{g}_d$ be a random Boolean function in variables $x_1, ..., x_d$. We are going to prove by induction on d that

$$\mathbb{E}\,[\mu(\mathbf{g}_d)] \leq d + 1 \tag{6}$$

Base. $d = 1$. Here we have $\mu(g(x_1)) \leq 2$ for *any* $g(x_1)$. This follows from (1) if $g = x_1^\epsilon$. By (4) and (1) we have $\mu(0) + \mu(1) \leq \mu(x_1) + \mu(\neg x_1) \leq 2$ which proves $\mu(g(x_1)) \leq 2$ in the remaining case, when g is a constant.

Inductive step. Assume that (6) is already proved for d. Let the symbol $\approx$ mean that two random functions are equally distributed. Note that

$$\mathbf{g}_{d+1} \approx \left(\mathbf{g}_d^0 \wedge x_{d+1}^0\right) \vee \left(\mathbf{g}_d^1 \wedge x_{d+1}^1\right) \tag{7}$$

where $\mathbf{g}_d^0$ and $\mathbf{g}_d^1$ are two independent copies of $\mathbf{g}_d$. By duality,

$$\mathbf{g}_{d+1} \approx \left(\mathbf{g}_d^0 \vee x_{d+1}^0\right) \wedge \left(\mathbf{g}_d^1 \vee x_{d+1}^1\right) \tag{8}$$

From (7) and (2) (remember that the latter is a consequence of (4)!) we have

$$\mathsf{E}\left[\mu(\mathbf{g}_{d+1})\right] \leq \mathsf{E}\left[\mu\left(\mathbf{g}_d^0 \wedge x_{d+1}^0\right)\right] + \mathsf{E}\left[\mu\left(\mathbf{g}_d^1 \wedge x_{d+1}^1\right)\right] \tag{9}$$

and similarly from (8) and (3),

$$\mathsf{E}\left[\mu(\mathbf{g}_{d+1})\right] \leq \mathsf{E}\left[\mu\left(\mathbf{g}_d^0 \vee x_{d+1}^0\right)\right] + \mathsf{E}\left[\mu\left(\mathbf{g}_d^1 \vee x_{d+1}^1\right)\right]. \tag{10}$$

Summing (9), (10) and applying consecutively (4), (1) and the inductive assumption (6), we obtain

$$
\begin{aligned}
2 \cdot \mathsf{E}\left[\mu(\mathbf{g}_{d+1})\right] \quad \leq \quad & \mathsf{E}\left[\mu\left(\mathbf{g}_d^0 \wedge x_{d+1}^0\right)\right] + \mathsf{E}\left[\mu\left(\mathbf{g}_d^0 \vee x_{d+1}^0\right)\right] + \\
& \mathsf{E}\left[\mu\left(\mathbf{g}_d^1 \wedge x_{d+1}^1\right)\right] + \mathsf{E}\left[\mu\left(\mathbf{g}_d^1 \vee x_{d+1}^1\right)\right] \\
\leq \quad & \mathsf{E}\left[\mu(\mathbf{g}_d^0)\right] + \mu(x_{d+1}^0) + \mathsf{E}\left[\mu(\mathbf{g}_d^1)\right] + \mu(x_{d+1}^1) \\
\leq \quad & 2 \cdot \mathsf{E}\left[\mu(\mathbf{g}_d)\right] + 2 \\
\leq \quad & 2d + 4.
\end{aligned}
$$

The inductive step is completed and (6) is proved.

Now the given function $f_n \in F_n$ can be expressed in the form

$$f_n = (\mathbf{g}_n \wedge (\mathbf{g}_n \oplus f_n \oplus 1)) \vee ((\mathbf{g}_n \oplus 1) \wedge (\mathbf{g}_n \oplus f_n)). \tag{11}$$

But $\mathbf{g}_n \approx \mathbf{g}_n \oplus f_n \oplus 1 \approx \mathbf{g}_n \oplus 1 \approx \mathbf{g}_n \oplus f_n$. So, applying to (11) the inequalities (2) and (3), averaging the result over $\mathbf{g}_n$ and applying (6) with $d = n$, we prove the desired bound $\mu(f_n) \leq O(n)$. $\square$

Theorems 1, 2 lead to the following result which shows the uselessness of Proposition 1 (unlike its monotone analogue!) for obtaining even superlinear lower bounds for the formula size over the standard basis with negations and resolves in the negative an open question from [13]:

Corollary 2 *For any $U, V \subseteq B^n$ such that $U \cap V = \varnothing$ and any non-zero matrix A over U, V (over an arbitrary field), the inequality*

$$\frac{\mathrm{rk}(A)}{\displaystyle\max_{R \in \mathcal{R}_{\mathrm{can}}(U,V)} \mathrm{rk}(A_R)} \leq O(n)$$

holds.

We conclude this paper with the following remark which is in a sense opposite to Corollary 2. Define a *partial matrix* over U, V to be an ordinary matrix over U, V with the exception that some entries can be left empty. The *rank* of a partial matrix A is the minimal rank of all possible full extensions of the partial matrix A. Proposition 1 can be strengthened by letting the matrix A be partial. Results contained in section 3 of the paper [13] imply that in this case the situation changes dramatically. Namely, the bound provided by the new version of Proposition 1 becomes almost universal in the context of graph complexity. If we prefer to stay in the Boolean framework, then we can claim (at least when the underlying field k is finite) that Proposition 1, applied to partial matrices A defined in the statement of Theorem 3.1 from [13], must provide exponential lower bounds for the formula size of almost all Boolean functions. Surely, the problem of getting actual lower bounds for $\mathrm{rk}(A)$ becomes extremely difficult in this context.

References

[1] Андреев А. Е. (1985) Об одном методе получения нижних оценок сложности индивидуальных монотонных функций, ДАН СССР, т. 282, № 5, с. 1033-1037. (Engl. transl. in: *Sov. Math. Dokl. 31, 530-534.*)

[2] Андреев А. Е. (1987) Об одном методе получения эффективных нижних оценок монотонной сложности, Алгебра и логика, т. 26, 1, с. 3-26.

[3] Разборов А. А. (1985) Нижние оценки монотонной сложности некоторых булевых функций, ДАН СССР, т. 281, № 4, с. 798-801. (Engl. transl. in: *Sov. Math. Dokl. 31, 354-357.*)

[4] Разборов А. А. (1985) Нижние оценки монотонной сложности логического перманента, Матем. зам., т. 37, вып. 6, с. 887-900. (Engl. transl. in: *Mathem. Notes of the Academy of Sci. of the USSR 37, 485-493.*)

[5] Разборов А. А. (1990) Нижние оценки сложности реализации симметрических булевых функций контактно-вентильными шемами, Матем. зам. т.48, вып. 6, с. 79-91.

[6] Храпченко В. М. (1971) О сложности реализации линейной функции в классе П-схем, Матем. зам., т. 9, вып. 1, с. 35-40. (Engl. transl. in: *Mathem. Notes of the Academy of Sci. of the USSR 11 (1972), 474-479.*)

[7] Alon N., Boppana R. B. (1987) The monotone circuit complexity of Boolean functions, Combinatorica, v. 7, 1, p. 1-22.

[8] Halsenberg B., Reischuk R. (1988) On Different Modes of Communication, Proc. 20th ACM STOC, p. 162-172.

[9] Karchmer M., Wigderson A. (1988) Monotone Circuits for Connectivity Require Super-logarithmic Depth, Proc. 20th ACM STOC, p. 539-550.

[10] Mehlhorn K., Schmidt E. M. (1982) Las Vegas is better than determinism in VLSI and distributive computing, Proc. 14th ACM STOC, p. 330-337.

[11] Raz R., Wigderson A. (1989) Probabilistic Communication Complexity of Boolean Relations, Proc. 30th IEEE FOCS, p. 562-567.

[12] Raz R., Wigderson A. (1990) Monotone Circuits for Matching Require Linear Depth, Proc. 22nd ACM STOC, p. 287-292.

[13] Razborov A. A. (1990) Applications of Matrix Methods to the Theory of Lower Bounds in Computational Complexity, Combinatorica v. 10, 1, p. 81-93.

[14] Razborov A. A. (1989) On the method of approximations, Proc. 21st ACM STOC, p. 167-176.

[15] Tárdos E. (1988) The gap between monotone and non-monotone circuit complexity is exponential, Combinatorica, v. 8, 1, p. 141-142.

[16] Wegener I. (1987) The Complexity of Boolean Functions, Wiley-Teubner.

Why is Boolean Complexity Theory Difficult?

Leslie G. Valiant[*]

1. Introduction

In the last decade substantial progress has been made in our understanding of restricted classes of Boolean circuits, in particular those restricted to have constant depth (Furst, Sipser, Saxe [FSS81], Ajtai [Ajt83], Yao [Yao85], Haiåstad [Has86], Razborov [Raz87], Smolensky [Smo87]) or to be monotone (Razborov [Raz85], Andreev [And85], Alon and Boppana [AB87], Tardos [Tar88], Karchmer and Wigderson [KW88]). The question arises, perhaps more urgently than before, as to what approaches could be pursued that might contribute to progress on the unrestricted model.

In this note we first argue that if P $\neq$ NP then any circuit-theoretic proof of this would have to be preceded by analogous results for the more constrained arithmetic model. This is because, as we shall observe, there are proven implications showing that if, for example, the Hamiltonian cycle problem (HC) requires exponential circuit size, then so does the analogous problem on arithmetic circuits. Since the set of valid algebraic identities in the latter model form a proper subset of those in the former, a lower bound proof for it should be strictly easier.

In spite of the above relationship the algebraic model is often regarded as an alternative, rather than a restriction of the Boolean model. One reason for this is that specific computations are usually understandable in one of these models, and not in both. In particular, the main power of the algebraic model derives from the possibility of cancellations, and it is usually difficult to express explicitly how these help in computing combinatorial problems. Our second aim in this note is to give an example of an algorithm, namely

---

[*]Aiken Computation Laboratory, Harvard University, Cambridge, MA 02138, and NEC Research Institute, Princeton, NJ 08540. Research at Harvard was supported by the National Science Foundation NSF-CCR-89-02500, the Office for Naval Research ONR-N0014-85-K-0445, the Center for Intelligent Control ARO DAAL 03-86-K-0171 and by DARPA AFOSR 89-0506.

the Samuelson-Berkowitz method for computing the determinant, where the intermediate terms that are computed but ultimately cancelled by the arithmetic circuit can be exhibited explicitly in combinatorial terms. The ease of computing the determinant can be attributed to the existence of such an auxiliary set of monomials with certain computational properties. A proof of an exponential lower bound on the complexity of a polynomial that is believed to be hard, such as the permanent or the Hamiltonian circuit polynomial, would involve establishing the nonexistence of such an auxiliary set. It is difficult to imagine how such a proof might go.

Finally, we observe that in low-level complexity, the arguments giving precedence to studying the algebraic model no longer hold. A major open problem area is that of proving for some explicit problem that it cannot be computed by unrestricted Boolean circuits simultaneously in size $O(n)$ and depth $O(\log n)$. We describe, via some conjectures, one candidate approach towards proving such a lower bound for problems such as sorting. Analogous conjectures exist for the algebraic model, but resolution of those would not imply the same for the Boolean case.

2. Algebraic Structures

Let $x_1, \cdots, x_n$ be a set of indeterminates and S a set of constants. Let $\otimes$ and $\oplus$ be two binary operators. We define a *circuit* or a *straight line program* syntactically as a finite sequence of instructions of the form:

$$f_i := g_i \; op_i \; h_i \qquad i \in \{1, \cdots, C\}$$

where, for each $i, op_i \in \{\otimes, \oplus\}$, and $g_i, h_i \in \{x_1, \cdots, x_n\} \cup S \cup \{f_1, \cdots, f_{i-1}\}$. In other words a circuit is a sequence of C binary instructions, where each argument of each one is either an indeterminate, a constant, or the result of the execution of an instruction earlier in the sequence. The complexity of a circuit is C, the number of instructions.

Such a circuit can be interpreted in several ways. In this paper we shall assume that S is a commutative ring with identity. Then each instruction f_i can be identified with the polynomial that is computed at f_i, if $\otimes$ and $\oplus$ are interpreted as the ring operations in the polynomial ring $S[x_1, \cdots, x_n]$.

Among natural multivariate polynomials whose complexity in this model is of interest are Hamiltonian circuits (HC), the permanent (PERM) and the determinant (DET). These are defined over a matrix X of indeterminates $\{x_{11}, \cdots, x_{nn}\}$ where x_{ij} can be thought of as representing edge (i, j) of the complete directed graph G_n on nodes $\{1, \cdots, n\}$. They are defined as follows:

$$PERM(X) \; = \; \sum_{\sigma \varepsilon \Sigma} \prod_{i=1}^{n} x_{i,\sigma(i)},$$

$$HC(X) \; = \; \sum_{\substack{\sigma \varepsilon \Sigma \\ \sigma \text{ has 1 cycle}}} \prod_{i=1}^{n} x_{i,\sigma(i)},$$

$$DET(X) \; = \; \sum_{\sigma \varepsilon \Sigma} (-1)^{\#(\sigma)} \prod_{i=1}^{n} x_{i,\sigma(i)},$$

where Σ is the set of permutations $\sigma : \{1, \cdots, n\} \to \{1, \cdots, n\}$ and $\#(\sigma)$ is the number of even length cycles in σ.

Clearly the monomials of PERM and DET correspond to cycle covers in G_n, while these of HC correspond to Hamiltonian circuits. These polynomials can be interpreted in any ring S. An important case is $S = \text{GF}[2]$, the finite field having two elements, in which case PERM = DET since then $-1 = 1$.

Each such problem is actually a polynomial family indexed by the number of indeterminates. When we talk about the permanent we mean the family $\text{PERM}_1, \text{PERM}_4, \text{PERM}_9, \cdots$ where the subscript denotes the number of variables. The computational complexity of such a family is determined by the family of circuits, one for each polynomial, that are the smallest circuits for each member. Thus PERM is polynomial time computable if and only if there is a family of circuits, of size growing polynomially in n, that computes PERM.

Different choices of S allow for different circuits.. For example PERM is polynomial time computable for $S=\text{GF}[2]$ but is not known to be so for any ring S whose characteristic is not a power of 2 (see [Val79]).

We shall define three complexity classes of families of polynomials in which all coefficients are integral multiples of unity: ARP is the class computable by polynomial size circuits in all rings, GFP is that computable by polynomial size circuits for $S = \text{GF}[2]$, and GFB is that computable by polynomial size circuits for $S=\text{GF}[2]$, where the algebra is assumed to obey the extra axiom $x^2 = x$. By definition, clearly,

$$ARP \subseteq GFP \subseteq GFB.$$

Now GFB is equivalent to the class of Boolean functions of polynomial circuit size, since the polynomial ring over $\text{GF}[2]$ with $x^2 = x$ gives Boolean algebra over $\{0, 1\}$ and the operations $\otimes$ and $\oplus$ in $\text{GF}[2]$ give a complete Boolean basis. Our observations here are first, that some natural polynomials in this algebra have natural combinatorial interpretations, and second, that their complexity can be related to NP-completeness.

In particular, a GFB circuit for HC computes nothing other than the parity of the number of Hamiltonian circuits in a graph. In other words if the indeterminates x_{ij} of such a circuit are set to 1 or 0 according to whether edge (i, j) is present in the graph then the circuit will compute 1 or 0 according to whether the number of Hamiltonian circuits is odd or even. Furthermore, the following is implicit in [VV86] since the randomized reductions used there can be simulated by small circuits [Adl78]:

Theorem 1 $HC \in GFB \Rightarrow NP$ *has polynomial size circuits.*

Denoting the class of Boolean functions of polynomial Boolean circuit size by pC [Val79], we conclude therefore that, since NP $\not\subseteq$ pC $\Rightarrow$ HC $\notin$ GFB, any efforts at proving the former should be retargeted at first proving the latter or indeed any of its even more restricted versions:

$$NP \not\subseteq pC \Rightarrow HC \notin GFB \Rightarrow HC \notin GFP \Rightarrow HC \notin ARP.$$

We note that $NP \not\subseteq pC$ is equivalent to the $P \neq NP$ question but with Turing uniformity removed from the definitions. It is argued in [Val79] that the nonuniform versions of such complexity questions as $P \neq NP$ are at least as natural as the uniform versions. For example, the nonuniform version of NP, there denoted by pD (for polynomial definable) has a completeness class that includes the original Hamiltonian circuits problem, just as NP has. Our conclusion, therefore, is that unless the explanation of the possible intractability of NP is to do with uniformity, one should seek it first in the more restricted structure of ARP or GFP.

3. Cancellations in the Samuelson-Berkowitz Algorithm

Any Boolean circuit can be re-expressed efficiently as a Boolean circuit over $\otimes$ and $\oplus$ in GF[2]. In almost any efficient such circuit the computation and cancellation of unwanted terms by the $\oplus$ operations plays a central role. Unfortunately, the way in which efficient computations exploit this facility is little understood. It is difficult even to find interesting examples where the structure of the cancelled terms can be exhibited explicitly.

In this section we study the one example we have found in a nontrivial computation where the terms that are computed but ultimately cancelled have a natural characterization. The algorithm is due to Samuelson [Sam42] and adapted by Berkowitz [Ber84] (see also Eberly [Ebe85]), and computes the determinant in ARP. (Note that other standard algorithms, such as Gaussian elimination, either use division or work only in certain rings.)

The algorithm for computing $\det(X)$ is as follows. We let B_k $(0 \le k \le n-1)$ be the principal $(n-k) \times (n-k)$ minor of X. Then we define an $(n-k) \times 1$ matrix C_k and a $1 \times (n-k)$ matrix D_k for each k $(1 \le k \le n-1)$ as follows

$$B_{k-1} = \begin{pmatrix} B_k & C_k \\ D_k & X_{n-k+1,n-k+1} \end{pmatrix}$$

where X_{ii} denotes the (i,i) element of matrix X.

For each k $(1 \le k \le n)$ we define T_k to be the $(n+2-k) \times (n+1-k)$ matrix defined as follows:

$$(T_k)_{ij} = \begin{cases} 0 & if \quad i > j+1 \\ -1 & if \quad i = j+1 \\ X_{n-k+1,n-k+1} & if \quad i = j \\ D_k B_k^{j-i-1} C_k & if \quad i < j \end{cases}$$

It turns out that the coefficients of the characteristic polynomial are given by the $(n+1) \times 1$ matrix

$$\prod_{k=1}^{n} T_k$$

where the (1,1) term gives the determinant.

The combinatorial interpretation of this computation of the determinant can be derived by noting that it is made up of the sum of terms of the form:

$$(T_1)_{1,i_1} (T_2)_{i_1,i_2} (T_3)_{i_2}, i_3 \cdots (T_{n-1})_{i_{n-2},i_{n-1}} (T_n)_{i_{n-1},1} \qquad (*)$$

Now the sequence of subscripts $1, i_1, i_2, \cdots i_{n-1}, 1$ imposes constraints on the products of matrix elements that occur in the term having this sequence of subscripts. From the definition of T_k it can be seen that $(T_k)_{ii} = X_{n-k+1,n-k+1}$ and $(T_k)_{ij} = D_k B_k^{j-i-1} C_k$ if $i < j$. Interpreted as paths in the underlying complete graph G_n, $(T_k)_{i,i}$ is a self-loop at node $n-k+1$, while $(T_k)_{ij}$ consists of closed walks of length $j-i+1$ going through only nodes $1, 2, \cdots, n-k+1$, and going through the last node $n-k+1$ exactly once. In both cases we can interpret $(T_k)_{ij}$ as a set of closed walks of length $j - i + 1$ since a self loop is of length one.

If s_k is the size of a closed walk generated by T_k, s_k being zero if $i_{k-1} = i_k + 1$, then clearly

$$s_k = i_k - i_{k-1} + 1.$$

Since $1 = i_0 = i_n = 1$ in $(*)$, it follows that

$$\sum_{1}^{n}(i_k - i_{k-1}) = \sum_{1}^{n}(s_k - 1) = 0.$$

Hence $\sum_{1}^{n} s_k = n$. Furthermore the sign of the term in $(*)$ is given by (-1) to the power of $p = \sum(s_k - 1)$ with summation over all k such that $s_k \geq 2$. Hence each even walk contributes a factor $(-1)^{s_k - 1} = -1$ and each odd walk a factor of $+1$. Therefore, the product $(*)$ contributes a factor

$$(-1)^{\# \text{ even length walks}}$$

which is exactly the same factor as in the determinant, if the only sets of walks counted are the sets of disjoint cycles.

To summarize, we define an m-loop in a graph with nodes $\{1, \cdots, n\}$ to be a closed walk going through node m exactly once, and through every $r > m$ zero times. A *loop-cover* of G is a set of m-loops, at most one for each m, such that the sum of the lengths of the loops is n. Note that a loop may repeat nodes and edges and that its length is the number of edges occurring in it allowing for multiplicity.

We claim that what the algorithm computes is the set of terms that correspond to loop covers of G_n, each term having sign $+$ or $-$ according to whether the loop cover has even or odd number of loops of even length.

Now a multi-set of n edges $\tilde{E}$ corresponds to more than one loop cover. For example the set of six edges shown below

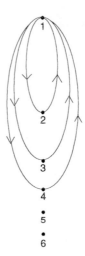

has six distinct loop covers. As can be verified, two of these correspond to single loops (of size 6), three correspond to two loops (of sizes 1 and 2 respectively in all cases) and one to three loops. Since all the loops involved have even length the algorithm will assign positive sign to the three covers with an even number of them, and negative sign to the three with an odd number of them. In other words these loop covers will cancel out.

Now the terms in the determinant correspond to loop covers that consist only of cycles that are disjoint. Clearly such sets of edges have only one loop cover, and the algorithm will give it the correct sign.

From the correctness of the algorithm and our combinatorial interpretation of the terms computed by it we conclude that

Theorem 2 *If $\tilde{E}$ is any multiset of n edges in G_n then*

1. *$\tilde{E}$ is a cycle cover $\Leftrightarrow$ $\tilde{E}$ forms exactly one loop cover,*

2. *$\tilde{E}$ is not a cycle cover $\Leftrightarrow$ $\tilde{E}$ forms an even number of loop covers, exactly a half of which have an odd number of even length loops.*

We note in conclusion that the Samuelson-Berkowitz algorithm is not multilinear, in the sense that it computes powers of the variables higher than one. It is an open problem as to whether a multilinear ARP algorithm exists for the determinant.

4. Simultaneous Lower Bounds on Size and Depth

We have argued that for NP-complete problems such as Hamiltonian circuits, super-polynomial lower bounds on Boolean circuits imply similar lower bounds on more restricted algebraic models of computation. Hence one would expect that resolving the latter problem should be mathematically more tractable than resolving the former. In low-level complexity, however, this formal relationship appears to break down, and there is fuller justification for working separately on both Boolean and algebraic problems. A case in question is that of simultaneous lower bounds on size and depth. For Boolean circuits it a major open problem to find an explicit family of sets of Boolean functions such that there is no family of Boolean circuits over a complete basis computing them that has size $O(n)$ and depth $O(\log n)$ simultaneously. Here n is the number of arguments plus functions. Below we shall give a combinatorial conjecture the truth of which would imply such lower bounds for problems including shifting and sorting. Analogous conjectures have been proposed in the algebraic context [Val77] and some progress made on them ([Fri90]).

Consider a bipartite graph G with node set $X \cup Y$ where $X = \{x_1, \cdots, x_n\}$ and $Y = \{y_1, \cdots, y_n\}$ denote input variables and output functions respectively. Suppose the edges are defined implicitly by a mapping τ where $\tau(y_i) \subseteq X$ is the set of input nodes that are adjacent to y_i in the graph.

Now we define m Boolean functions $f_1(\underline{x}), \cdots, f_m(\underline{x})$ and n further Boolean functions $g_1, \cdots, g_n$, where g_i has $m + |\tau(y_i)|$ Boolean arguments, such that, with some abuse of notation, $y_i = g_i(f_1(\underline{x}), \cdots, f_m(\underline{x}), \tau(y_i))$. In other words we have m functions of the inputs, and each output y_i can be an arbitrary function of these m common bits and of the inputs $\tau(y_i)$ to which it is connected directly in G.

We say that G realizes permutation σ with m common bits if there exist $f_1, \cdots, f_m, g_1, \cdots, g_n$ such that for all i, $1 \leq i \leq n$,

$$x_{\sigma(i)} = g_i(f_1(\underline{x}), \cdots, f_m(\underline{x}), \tau(y_i)).$$

In other words, for the fixed G and given σ one can find the appropriate Boolean functions $\{g_i\}, \{f_j\}$ such that the $\{y_i\}$ realize the permutation σ of $\{x_i\}$, for all truth assignments to the $\{x_i\}$.

Now it appears that if G is sparse, say with degree less than $n^{1-\varepsilon}$ for some $\varepsilon > 0$, and if m is small enough (e.g. $m < n/2$) then the common bits form an information bottleneck, and G cannot realize all permutations. In particular:

Conjecture 1 *If G has degree 3 then there is a permutation σ such that G does not realize σ with $n/2$ common bits.*

Conjecture 2 *If G has degree 3 then there is a cyclic shift σ such that G does not realize σ with $n/2$ common bits.*

Conjecture 1′ and 2′ *Conjectures 1 and 2 but for G with $O(n^{1+\varepsilon})$ edges for some $\varepsilon > 0$.*

Now Conjecture 2′ would imply that the following shifting problem has no circuit of size $O(n)$ and depth $O(\log n)$. The shifting problem is defined with $n + \lceil \log_2 n \rceil$ inputs $x_0, \cdots, x_{n-1}, s_1, \cdots, s_{\lceil \log_2 n \rceil}$ where the value of $\underline{s}$ defines in binary the amount s to be shifted. In other words output y_i equals $x_{i-s \bmod n}$ where s is the shift.

To prove the implication we appeal to Proposition 6.2 in [Val77] which implies that for any $\varepsilon > 0$, in any n-input n-output graph family of size $O(n)$ and depth $O(\log n)$ some $n/2$ nodes and adjacent edges can be removed so that fewer than $O(n^{1+\varepsilon})$ input output pairs remain connected. Hence if such a computation graph existed for shifting (i.e. if we identified $x_1, \cdots, x_n$ of the shifter as the inputs of the graph) then we could identify the $n/2$ removed nodes as the common bits. Each setting of the shift bits $\{s_i\}$ could be seen as setting the functions $\{f_i\}$ and $\{g_i\}$ so as to contradict Conjecture 2′.

Conjecture 1′ is a weaker statement than Conjecture 2′, and might be easier to prove. Potentially it could yield a lower bound on sorting, say of $n/(2 \log n)$ numbers each of size $2 \log n$. The conjecture as stated is not quite enough but the reader can verify that it can be adapted in various ways so as to imply such a lower bound.

As our conjectures imply, little is known about the permutations that can be realized even for low degrees such as three. The degree one case can be analyzed completely. The canonical case is the identity graph $G(\tau(y_i) = x_i)$ which, with $n/2$ common bits, can realize such permutations as $y_{2i} = x_{2i-1}, y_{2i-1} = x_{2i}$ by letting $f_i = x_{2i-1}, y_{2i} = f_i \oplus x_{2i}, y_{2i-1} = f_i \oplus x_{2i-1}$.

The degree two case has been analysed partially by K. Kalorkoti.

References

[AB87]   N. Alon and R.B. Boppana. The monotone circuit complexity of Boolean functions. *Combinatorica*, 7(1):1–22, (1987).

[Adl78] L. Adleman. Two theorems on random polynomial time. In *Proc. 19th IEEE Symp. on Foundations of Computer Science*, pages 75–83, (1978).

[Ajt83] M. Ajtai. $\sum_1^1$-formulae on finite structures. *Annals of Pure and Applied Logic*, 24:1–48, (1983).

[And85] A.E. Andreev. On a method for obtaining lower bounds for the complexity of individual monotone funtions. *Sov. Math. Dokl*, 31(3):530–534, (1985).

[Ber84] S.J. Berkowitz. On computing the determinant in small parallel time using a small number of processors. *Inf. Proc. Letters*, 18:147–150, (1984).

[Ebe85] W. Eberly. Very fast parallel matrix and polynomial arithmetic. Technical Report 178/85, Comp. Sci. Dept., Univ. of Toronto, (1985).

[Fri90] J. Friedman. A note on matrix rigidity. Manuscript, (1990).

[FSS81] M. Furst, J.B. Saxe, and M. Sipser. Parity circuits and the polynomial time hierarchy. In *Proc. 22nd IEEE Symp. on Foundations of Computer Science*, pages 260–270, (1981). Also Math. Systems Theory 17, (1984) 13-27.

[Has86] J. Håstad. Improved lower bounds for small depth circuits. In *Proc. 18th ACM Symp. on Theory of Computing*, pages 6–20, (1986). Also Computational Limitations for Small Depth Circuits, (1988), MIT Press, Cambridge, MA, USA.

[KW88] M. Karchmer and A Wigderson. Monotone circuits for connectivity require super-logarithmic depth. In *Proc. 20th ACM Symp. on Theory of Computing*, pages 539–550, (1988).

[Raz85] A.A. Razborov. Lower bounds on the monotone complexity of some boolean functions. *Sov. Math. Dokl.*, 31:354–357, (1985).

[Raz87] A.A Razborov. Lower bounds for the size of circuits for bounded depth with basis $\{\wedge, \oplus\}$. *Math Notes of the Acad. of Sciences of the USSR*, 41(4):333–338, (1987).

[Sam42] P.A. Samuelson. A method of determining explicitly the coefficients of the characteristic equation. *Ann. Math. Statist.*, 13:424–429, (1942).

[Smo87] R. Smolensky. Algebraic methods in the theory of lower bounds for boolean circuit complexity. In *19th ACM Symp. on Theory of Computing*, pages 77–82, (1987).

[Tar88] E. Tardos. The gap between monotone and non-monotone circuit complexity is exponential. *Combinatorica*, 8(1):141–142, (1988).

[Val77] L.G. Valiant. Graph-theoretic arguments in low-level complexity. *Lecture Notes in Compter Science*, 53:162–176, Springer (1977).

[Val79] L.G. Valiant. Completeness classes in algebra. In *Proc. 11th ACM Symp. on Theory of Computing*, pages 249–261, (1979).

[VV86] L.G. Valiant and V.V. Vazirani. NP is as easy as detecting unique solutions. *Theoretical Computer Science*, 47:85–93, (1986).

[Yao85] A.C.-C. Yao. Separating the polynomial time hierarchy by oracles. In *Proc. 26th IEEE Symp. on Foundations of Computer Science*, pages 1–10, (1985).

The Multiplicative Complexity of Boolean Quadratic Forms, a Survey.

Roland Mirwald[*] *Claus P. Schnorr*[*]

Abstract

In this survey we present lower and upper bounds on the multiplicative complexity $L(f_A)$ and $L(f_A, f_B)$ of Boolean quadratic forms f_A and f_B represented by Boolean matrices A and B. We give explicit formulae that determine $L(f_A, f_B)$. We also evaluate the rank $R(A, B)$ of a pair of Boolean matrices A and B. The main new results are that $L(f_A, f_B) = \frac{1}{2}R(A \oplus A^\mathsf{T}, B \oplus B^\mathsf{T})$ and $R(A, B) = \lceil \frac{1}{2}(R(A) + R(B) + R(A \oplus B) + d) \rceil$ where d is a function of the number of exceptional blocks and the number of unsaturated blocks in the Kronecker normal form of the pair (A, B). These formulae hold for all pairs (A, B) with some exceptions.

1. Introduction

A function $f : \{0,1\}^n \to \{0,1\}$ is called a *Boolean function* in n variables. Let B_n be the set of all Boolean functions in n variables. B_n is a ring under the exclusive or $(\oplus, \mathrm{XOR})$ as addition and the conjunction $(\wedge, \mathrm{AND})$ as multiplication.

Let the elements of the Galois field $\mathbb{Z}_2 = \mathbb{Z}/2\mathbb{Z}$ be 0 and 1. Any polynomial $p \in \mathbb{Z}_2[x_1, \ldots, x_n]$ defines a function $\varphi(p) : \{0,1\}^n \to \{0,1\}$. The mapping

$$\varphi : \mathbb{Z}_2[x_1, \ldots, x_n] \to \mathrm{B}_n$$

[*]Universität Frankfurt, Fachbereich Mathematik/Informatik, Postfach 111932, 6000 Frankfurt am Main, Germany.

is a ring homomorphism which is surjective. The kernel of φ is the ideal $I \subset \mathbb{Z}_2[x_1, \ldots, x_n]$ generated by the polynomials $x_1 + x_1^2, \ldots, x_n + x_n^2$. Thus we have a ring isomorphism

$$\mathbb{Z}_2[x_1, \ldots, x_n] / I \cong \mathrm{B}_n.$$

The ring $\mathbb{Z}_2[x_1, \ldots, x_n] / I$ is called the ring of *Boolean polynomials* in n Boolean variables. The residue class $x_i + I$ is called a *Boolean variable*. We identify $\mathbb{Z}_2[x_1, \ldots, x_n] / I$ with B_n and we abbreviate $x_i + I$ as x_i.

We will use the following linear subspaces of B_n:

- $\mathrm{B}_{n,1} = \langle x_1, \ldots, x_n \rangle$, the space of Boolean linear forms,

- $\mathrm{B}_{n,2} = \langle x_i x_j \mid 1 \leq i < j \leq n \rangle$, the space of Boolean quadratic forms.

Here we let $\langle f_1, \ldots, f_r \rangle \subset \mathrm{B}_n$ denote the linear space that is generated by $f_1, \ldots, f_r$, i.e. the span of $f_1, \ldots, f_r$. In this paper all quadratic and linear forms are Boolean. We write $f = g \,(\mathrm{mod}\,\mathrm{B}_{n,1})$ if $f \oplus g \in \mathrm{B}_{n,1}$. In order to abbreviate formulae we suppress all $\wedge$–symbols.

Definition (multiplicative complexity, level–one multiplicative complexity)

(1) The *multiplicative complexity* $\mathrm{L}(f_1, \ldots, f_s)$ of a set of Boolean polynomials $f_1, \ldots, f_s \in \mathrm{B}_n$ is the minimal number $r \in \mathbb{N}$ for which there exist Boolean polynomials $u_i, v_i, g_i \in \mathrm{B}_n$ for $i = 1, \ldots, r$ such that

   (i) $u_i, v_i \in \langle g_1, \ldots, g_{i-1}, \ x_1, \ldots, x_n, \ 1 \rangle$, $g_i = u_i v_i$ for $i = 1, \ldots, r$,

   (ii) $f_1, \ldots, f_s \in \langle g_1, \ldots, g_r, \ x_1, \ldots, x_n, \ 1 \rangle$.

(2) The *level–one multiplicative complexity* $\mathrm{L}_1(f_1, \ldots, f_s)$ of a set of Boolean quadratic forms $f_1, \ldots, f_s \in \mathrm{B}_{n,2}$ is the minimal number $r \in \mathbb{N}$ for which there exist linear forms $u_i, v_i \in \mathrm{B}_{n,1}$ for $i = 1, \ldots, r$ such that

$$f_1, \ldots, f_s \in \langle u_1 v_1, \ldots, u_r v_r, \ x_1, \ldots, x_n \rangle.$$

The multiplicative complexity $\mathrm{L}(f_1, \ldots, f_s)$ is the minimal number of $\wedge$–gates in any Boolean circuit that computes $f_1, \ldots, f_s$ using merely $\wedge$–gates and $\oplus$–gates ($\oplus$–gates are for free). The level–one multiplicative complexity $\mathrm{L}_1(f_1, \ldots, f_s)$ is the minimal number of $\wedge$–gates in any circuit that computes the quadratic forms $f_1, \ldots, f_s \in \mathrm{B}_{n,2}$ using merely $\wedge$–gates and $\oplus$ gates with only one level of $\wedge$–gates.

Clearly we have

$$\mathrm{L}(f_1, \ldots, f_s) \leq \mathrm{L}_1(f_1, \ldots, f_s)$$

for all $f_1, \ldots, f_s \in B_{n,2}$. The complexity measures L and L_1 coincide for single quadratic forms and for pairs of quadratic forms (see Theorems 3 and 8).

In the next section we characterize the multiplicative complexity of a single quadratic form f. By Theorem 2 the multiplicative complexity $L(f)$ is half the rank of an associated matrix. Theorem 3 implies that $L(f) = L_1(f)$ for all $f \in B_{n,2}$.

In Section 3 we relate the level–one multiplicative complexity of sets of quadratic forms to the rank of an associated set of Boolean matrices (Theorem 4). By Theorem 5 there exists for all pairs $f_1, f_2 \in B_{n,2}$ a *common part*, i.e. a quadratic form $g \in B_{n,2}$ with $L(f_i) = L(f_i \oplus g) + L(g)$ for $i = 1, 2$ and $L_1(f_1, f_2) = L(f_1 \oplus g) + L(f_2 \oplus g) + L(g)$. As a consequence we prove the lower bound $L(f_1, f_2) \geq \lceil \frac{1}{2}((L(f_1) + L(f_2) + L(f_1 \oplus f_2)) \rceil$ for all $f_1, f_2 \in B_{n,2}$. Theorem 5 is also the key for studying computational independence of $f_1, f_2 \in B_{n,2}$ and for proving that $L(f_1, f_2) = L_1(f_1, f_2)$ holds for all $f_1, f_2 \in B_{n,2}$ (Theorem 8).

Section 4 contains lower and upper bounds on the multiplicative complexity of pairs of quadratic forms f_1, f_2. It turns out that $L(f_1, f_2)$ is closely related to the rank of an associated pair of Boolean matrices. Therefore we also study the rank of pairs of matrices over the field $\mathbb{Z}_2$. The rank of matrix pairs over algebraically closed fields and over fields that contain "enough" elements has been characterized by GRIGORYEV (1978) and JA' JA' (1979). HÅSTAD (1990) has shown that the problem of deciding the rank of an arbitrary set of matrices over a finite field is NP–complete.

2.   The Multiplicative Complexity of Single Boolean Quadratic Forms

Let $M_{m,n}(\mathbb{Z}_2)$ denote the set of $m \times n$ matrices with entries from $\mathbb{Z}_2$. Boolean quadratic forms in $B_{n,2}$ can be represented by matrices $A = (a_{ij})_{1 \leq i,j \leq n} \in M_{n,n}(\mathbb{Z}_2)$. We associate with the matrix A the Boolean quadratic form $f_A \in B_{n,2}$ defined by

$$f_A = \sum_{i \neq j} a_{ij} x_i x_j.$$

The mapping

$$M_{n,n}(\mathbb{Z}_2) \rightarrow B_{n,2}$$
$$A \mapsto f_A$$

is a homomorphism of linear spaces. This homomorphism is surjective, its kernel is the linear space of skew symmetric matrices. This implies

Fact 1 *For all* $A, B \in M_{n,n}(\mathbb{Z}_2)$ *the following conditions are equivalent:*

(1) $f_A = f_B$.

(2) $A \oplus B$ *is skew symmetric.*

(3) $A \oplus A^\top = B \oplus B^\top$.

We let $A^\top$ denote the transpose of the matrix A. According to Fact 1 we have a bijection

$$B_{n,2} \;\to\; M_{n,n}(\mathbb{Z}_2)$$
$$f_A \;\mapsto\; A \oplus A^\top .$$

By the following theorem the multiplicative complexity of any quadratic form f_A is half the rank of the matrix $A \oplus A^\top$. The matrix $A \oplus A^\top$ is *alternate*, i.e. it is skew symmetric with zero diagonal. It is well known that alternate matrices have even rank. We let $R(B)$ denote the rank of the matrix $B \in M_{n,n}(\mathbb{Z}_2)$.

Theorem 2 (MIRWALD, SCHNORR (1987), Theorem 3.6)
For all $f_A \in B_{n,2}$ *we have* $L(f_A) = \frac{1}{2} R(A \oplus A^\top)$.

The proof of Theorem 2 proceeds by showing first that $L_1(f_A) = \frac{1}{2} R(A \oplus A^\top)$ for all quadratic forms f_A. We call a representation

$$f = \bigoplus_{i=1}^{t} u_i v_i \;(\mathrm{mod}\,B_{n,1}) \tag{1}$$

of a quadratic form f with linear forms u_i, v_i *minimal* if $t = L_1(f)$. It is shown that a representation (1) is minimal if and only if the linear forms $u_1, \ldots, u_t, v_1, \ldots, v_t$ are linearly independent. In a second step we have shown that the complexity measures L and L_1 coincide for single quadratic forms. More precisely, all circuits that compute $f \in B_{n,2}$ and have more than one level of $\wedge$–gates are not minimal: they need at least $L(f) + 1$ many $\wedge$–gates.

Theorem 3 (MIRWALD, SCHNORR (1987), Theorem 3.5)
Every circuit that computes a quadratic form f using $L(f)$ many $\wedge$–gates has at most one level of $\wedge$–gates.

We identify linear forms $u = \bigoplus_{i=1}^{n} a_i x_i \in B_{n,1}$ with the corresponding row vector $(a_1, \ldots, a_n) \in M_{1,n}(\mathbb{Z}_2)$. The *tensor product* $u \otimes v$ of two row vectors $u, v \in M_{1,n}(\mathbb{Z}_2)$ is the rank–1 matrix $u \otimes v = u^\mathsf{T} v$. Let f be a quadratic form and let u_i, v_i for $i = 1, \ldots, t$ be linear forms. Then the following statements are equivalent:

- $f = \bigoplus_{i=1}^{t} u_i v_i \pmod{B_{n,1}}$

- $f = f_A \iff A \oplus A^\mathsf{T} = \bigoplus_{i=1}^{t} (u_i \otimes v_i \oplus v_i \otimes u_i)$.

We define the *dual domain* $\mathrm{D}(f_A)$ of a quadratic form f_A as the subspace of $B_{n,1}$ that corresponds to the linear space generated by the row vectors of the matrix $A \oplus A^\mathsf{T}$. Now Theorem 2 implies that for all quadratic forms f we have

$$\dim \mathrm{D}(f) = 2\,\mathrm{L}(f).$$

Moreover it follows from Theorem 2 that for any representation

$$f = \bigoplus_{i=1}^{t} u_i v_i \pmod{B_{n,1}}$$

of a quadratic form f with linear forms u_i, v_i we have that $\mathrm{D}(f) \subset \langle u_1, \ldots, u_t, v_1, \ldots, v_t \rangle$ and we have $\mathrm{D}(f) = \langle u_1, \ldots, u_t, v_1, \ldots, v_t \rangle$ iff the representation is minimal, i.e. iff $t = \mathrm{L}(f)$.

3. Independence and Lower Bounds for Two Boolean Quadratic Forms

We now establish the relation between the multiplicative complexity of Boolean quadratic forms and the rank of sets of Boolean matrices.

Definition (rank of a set of matrices)
The *rank* $\mathrm{R}(A_1, \ldots, A_s)$ of a set of matrices $A_1, \ldots, A_s \in M_{m,n}(\mathbb{Z}_2)$ is the minimal number $r \in \mathbb{N}$ with the property that there exist matrices $P_1, \ldots, P_r$ with rank 1 such that $A_1, \ldots, A_s \in \langle P_1, \ldots, P_r \rangle$.

In the situation of an arbitrary field K and quadratic forms in $\mathrm{K}[x_1, \ldots, x_n]$ STRASSEN (1973) has related the multiplicative complexity of sets of quadratic forms to the rank of a corresponding set of matrices. The following theorem adjusts Strassen's result to Boolean quadratic forms.

Theorem 4 (MIRWALD, SCHNORR (1987), Theorem 5.2)
Let $A_1, \ldots, A_s$ be matrices in $M_{n,n}(\mathbb{Z}_2)$ and let $f_{A_1}, \ldots, f_{A_s} \in B_{n,2}$ be the corresponding Boolean quadratic forms. Then we have

(1) $L_1(f_{A_1}, \ldots, f_{A_s}) =$
   $\min\{R(A_1 \oplus S_1, \ldots, A_s \oplus S_s) \mid S_1, \ldots, S_s \in M_{n,n}(\mathbb{Z}_2) \text{ skew symmetric}\}$,

(2) $\frac{1}{2} R(A_1 \oplus A_1^\mathsf{T}, \ldots, A_s \oplus A_s^\mathsf{T}) \leq L_1(f_{A_1}, \ldots, f_{A_s}) \leq R(A_1, \ldots, A_s)$.

We know from Theorem 2 that the lower bound

$$\lceil \tfrac{1}{2} R(A_1 \oplus A_1^\mathsf{T}, \ldots, A_s \oplus A_s^\mathsf{T}) \rceil \leq L_1(f_{A_1}, \ldots, f_{A_s})$$

in Theorem 4 (2) is actually an equality for $s = 1$. In Section 4 we will see that equality also holds for $s = 2$ in all cases where we can determine $R(A_1 \oplus A_1^\mathsf{T}, A_2 \oplus A_2^\mathsf{T})$ by explicit formulae.

The following theorem plays an important role.

Theorem 5 (MIRWALD, SCHNORR (1987), Theorem 4.1)
For all quadratic forms $f_1, f_2 \in B_{n,2}$ there exists a quadratic form $g \in B_{n,2}$ such that

(1) $L(f_i) = L(f_i \oplus g) + L(g)$ *for $i = 1, 2$,*

(2) $L_1(f_1, f_2) = L(f_1 \oplus g) + L(f_2 \oplus g) + L(g)$.

We call a quadratic form g satisfying properties (1) and (2) in Theorem 5 a *common part* for f_1 and f_2. Theorem 5 means that for all $f_1, f_2 \in B_{n,2}$ there exists an L_1–minimal circuit that contains L–minimal circuits for both f_1 and f_2. The existence of a common part for f_1 and f_2 implies a lower bound on $L_1(f_1, f_2)$:

Corollary 6 *For all quadratic forms $f_1, f_2 \in B_{n,2}$ we have*

$$L_1(f_1, f_2) \geq \lceil \tfrac{1}{2} (L(f_1) + L(f_2) + L(f_1 \oplus f_2)) \rceil.$$

Proof. Let $g \in B_{n,2}$ be as in Theorem 5. Then we have

$$\begin{aligned}
L_1(f_1, f_2) &= L(f_1 \oplus g) + L(f_2 \oplus g) + L(g) \\
&= L(f_1) + L(f_2) - L(g) \\
&\geq L(f_1 \oplus f_2) + L(g).
\end{aligned}$$

Hence $2\,L_1(f_1, f_2) \geq L(f_1) + L(f_2) + L(f_1 \oplus f_2)$. $\qquad\qquad\square$

Two quadratic forms $f_1, f_2 \in B_{n,2}$ are called *computationally independent* if $L_1(f_1, f_2) = L(f_1) + L(f_2)$. It follows from Theorem 5 that f_1, f_2 are computationally independent iff the zero quadratic form is a common part for f_1 and f_2.

Let f be a quadratic form and let U be a linear subspace of the dual domain $D(f)$. Then f is called *non linear* on U if there exist $u, v \in U$ with $L(f \oplus uv) = L(f) - 1$, otherwise f is called *linear* on U.

Theorem 7 (MIRWALD, SCHNORR (1987), Theorem 4.3)
Let $f \in B_{n,2}$ with $L(f) = t$ and let $U \subset D(f)$ be a linear subspace with $\dim U = r$. Then f is linear on U if and only if there is a minimal representation $f = \bigoplus_{i=1}^{t} u_i v_i \pmod{B_{n,1}}$ for f such that $u_1, \ldots, u_r$ is a basis for U.

As a consequence of Theorem 7 MIRWALD and SCHNORR (1987) have presented a polynomial time algorithm which decides whether $f_1, f_2 \in B_{n,2}$ are computationally independent.

By Theorem 7 the quadratic forms

$$\begin{aligned}
f_1 &= x_1 x_2 \oplus x_3 x_4 \oplus \cdots \oplus x_{4t-1} x_{4t}, \\
f_2 &= x_1 x_3 \oplus x_5 x_7 \oplus \cdots \oplus x_{4t-3} x_{4t-1}
\end{aligned}$$

are computationally independent since f_1 depends linearly on the variables of f_2. Hence we have

$$L_1(f_1, f_2) = L(f_1) + L(f_2) = 3\,t.$$

It is interesting to note that the lower bound of Corollary 6 is rather weak for the case of two quadratic forms that are computationally independent. For the above quadratic forms f_1, f_2, Corollary 6 merely implies that $L_1(f_1, f_2) \geq 2.5\,t$. On the other hand the lower bound of Corollary 6 is tight for all pairs $f_1, f_2 \in B_{n,2}$ that have a common part g such that $L(g)$ is maximal for all $f_1, f_2 \in B_{n,2}$. The complexity of the above quadratic forms $f_1, f_2 \in B_{4t,2}$ is maximal for all pairs of quadratic forms in $B_{4t,2}$.

Another consequence of Theorems 5 and 7 is that the complexity measures L and L_1 coincide for pairs of quadratic forms.

Theorem 8 (MIRWALD, SCHNORR (1987), Theorem 4.6)
For all $f_1, f_2 \in B_{n,2}$ we have $L(f_1, f_2) = L_1(f_1, f_2)$.

We finally note that Theorem 5 and Corollary 6 have a counterpart for Boolean matrices:

Theorem 9 (MIRWALD (1991))
For all matrices $A_1, A_2 \in M_{m,n}(\mathbb{Z}_2)$ there exists a matrix $G \in M_{m,n}(\mathbb{Z}_2)$ such that

(1) $R(A_i) = R(A_i \oplus G) + R(G)$ *for* $i = 1, 2$,

(2) $R(A_1, A_2) = R(A_1 \oplus G) + R(A_2 \oplus G) + R(G)$.

Theorem 9 yields a lower bound for the rank of pairs of Boolean matrices corresponding to the lower bound of Corollary 6:

$$R(A_1, A_2) \geq \lceil \tfrac{1}{2}\left(R(A_1) + R(A_2) + R(A_1 + A_2)\right)\rceil.$$

4.   The Multiplicative Complexity of Pairs of Quadratic Boolean Forms

We introduce isomorphism classes of pairs (f_A, f_B) of Boolean quadratic forms. They correspond to congruence classes of the matrix pairs $(A \oplus A^\mathsf{T}, B \oplus B^\mathsf{T})$. We determine the complexity $L(f_A, f_B)$ in terms of invariants related to the Kronecker normal form of the pair $(A \oplus A^\mathsf{T}, B \oplus B^\mathsf{T})$. For this purpose we also study the rank of pairs of arbitrary matrices over $\mathbb{Z}_2$.

Definition Two pairs $(A_1, B_1), (A_2, B_2) \in M_{n,n}(\mathbb{Z}_2)^2$ are called *congruent* if there exists a regular (i.e. invertible) matrix $T \in M_{n,n}(\mathbb{Z}_2)$ such that $T A_1 T^\mathsf{T} = A_2$ and $T B_1 T^\mathsf{T} = B_2$.

The congruence defines an equivalence relation for pairs $(A, B) \in M_{n,n}(\mathbb{Z}_2)^2$ which preserves the complexity $L(f_A, f_B)$ and the rank $R(A, B)$. More precisely for regular $T \in M_{n,n}(\mathbb{Z}_2)$ and arbitrary $A, B \in M_{n,n}(\mathbb{Z}_2)$ we have

$$\begin{aligned} L(f_A, f_B) &= L(f_{T A T^\mathsf{T}}, f_{T B T^\mathsf{T}}), \\ R(A, B) &= R(T A T^\mathsf{T}, T B T^\mathsf{T}). \end{aligned}$$

For example
$$A, B \in \langle u_1 \otimes v_1, \ldots, u_t \otimes v_t \rangle$$
holds if and only if

$$TAT^\top, \; TBT^\top \in \langle (u_1T) \otimes (v_1T), \ldots, (u_tT) \otimes (v_tT) \rangle.$$

Definition Two pairs of quadratic forms (f_{A_1}, f_{B_1}), $(f_{A_2}, f_{B_2}) \in (B_{n,2})^2$ are called *isomorphic* if there exists a regular matrix $T \in M_{n,n}(\mathbb{Z}_2)$ such that $f_{TA_1T^\top} = f_{A_2}$ and $f_{TB_1T^\top} = f_{B_2}$.

Isomorphic pairs have the same multiplicative complexity.

Next we consider congruence classes of alternate matrix pairs $(A \oplus A^\top, \; B \oplus B^\top)$. The motivation for this is that two pairs $(f_{A_i}, \; f_{B_i}) \in B_{n,2}$ for $i = 1, 2$ are isomorphic if and only if the matrix pairs $(A_i \oplus A_i^\top, \; B_i \oplus B_i^\top)$ for $i = 1, 2$ are congruent. The next theorem establishes a particularly nice representative in the congruence class of an alternate matrix pair. We associate with an arbitrary matrix $C \in M_{r,s}(\mathbb{Z}_2)$ the matrix

$$\tilde{C} = \begin{bmatrix} 0 & C \\ 0 & 0 \end{bmatrix} \in M_{r+s,r+s}(\mathbb{Z}_2).$$

Theorem 10 (MIRWALD (1991))
Let $(A, B) \in M_{n,n}(\mathbb{Z}_2)^2$ be an arbitrary pair of alternate matrices. Then there exist matrices $C, D \in M_{r,s}(\mathbb{Z}_2)$ with $r + s = n$ such that (A, B) is congruent to the pair

$$(\tilde{C} \oplus \tilde{C}^\top, \; \tilde{D} \oplus \tilde{D}^\top) = \left(\begin{bmatrix} 0 & C \\ C^\top & 0 \end{bmatrix}, \begin{bmatrix} 0 & D \\ D^\top & 0 \end{bmatrix} \right).$$

The congruence of pairs of alternate matrices $A \oplus A^\top$, $B \oplus B^\top$ is best understood through the following equivalence relation for pairs of arbitrary matrices.

Definition Two pairs (A_1, B_1), $(A_2, B_2) \in M_{m,n}(\mathbb{Z}_2)^2$ are called *equivalent* if there exist regular matrices $S \in M_{m,m}(\mathbb{Z}_2)$, $T \in M_{n,n}(\mathbb{Z}_2)$ such that $SA_1T = A_2$ and $SB_1T = B_2$.

We note that equivalent pairs of matrices have the same rank. Mirwald shows the following theorem in his dissertation.

Theorem 11 (MIRWALD (1991))
Two pairs of alternate matrices are congruent if and only if they are equivalent.

The equivalence class of a pair $(A, B) \in M_{m,n}(\mathbb{Z}_2)^2$ is characterized by the Kronecker normal form of the pair which we introduce next.

Let $\mathrm{diag}(A_1, \ldots, A_s)$ denote the matrix that has the matrices $A_1, \ldots, A_s$ in its diagonal:

$$\mathrm{diag}(A_1, \ldots, A_s) = \begin{bmatrix} A_1 & & & \\ & A_2 & & \mathbf{0} \\ & & \ddots & \\ & \mathbf{0} & & A_s \end{bmatrix}.$$

Here $A_1, \ldots, A_s$ are arbitrary rectangular matrices. Let E_n denote the $n \times n$ identity matrix.

Definition (Kronecker normal form, blocks and types of blocks)
A pair of matrices $(A, B) \in M_{m,n}(\mathbb{Z}_2)^2$ is in *Kronecker normal form* if there exist matrices $A_i, B_i \in M_{m_i, n_i}(\mathbb{Z}_2)$ for $i = 0, \ldots, s$ such that

$$A = \mathrm{diag}(A_0, \ldots, A_s), \quad B = \mathrm{diag}(B_0, \ldots, B_s),$$

where $A_0, B_0 \in M_{m_0, n_0}(\mathbb{Z}_2)$ are zero–matrices with $0 \le m_0 \le m$, $0 \le n_0 \le n$, and for $i = 1, \ldots, s$ each pair $(A_i, B_i) \in M_{m_i, n_i}(\mathbb{Z}_2)^2$ is a *block* of one of the following five types of blocks:

- *Blocks (A_i, B_i) of type 1:* $m_i = n_i \ge 1$ and

$$A_i = \begin{bmatrix} 0 & & & 1 \\ 1 & \ddots & & \alpha_1 \\ & \ddots & \ddots & \vdots \\ & & 1 & \alpha_{n_i - 1} \end{bmatrix}, \quad B_i = E_{n_i},$$

where the polynomial $1 + \alpha_1 \lambda + \cdots + \alpha_{n-1} \lambda^{n_i - 1} + \lambda^{n_i} \in \mathbb{Z}_2[\lambda]$ is a power of some irreducible polynomial in $\mathbb{Z}_2[\lambda]$. In case $m_i = n_i = 1$ we have $A_i = [1]$.

- *Blocks (A_i, B_i) of type 2:* $m_i = n_i \ge 1$ and

$$A_i = \begin{bmatrix} 0 & & \\ 1 & \ddots & \\ & \ddots & \ddots \\ & & 1 & 0 \end{bmatrix}, \quad B_i = E_{n_i}.$$

In case $m_i = n_i = 1$ we have $A_i = [0]$.

- *Blocks* (A_i, B_i) *of type* 3: $m_i = n_i \geq 1$ and (B_i, A_i) is a block of type 2.

- *Blocks* (A_i, B_i) *of type* 4: $m_i = n_i - 1 \geq 1$ and

$$A_i = \begin{bmatrix} 0 & 1 & & \\ & \ddots & \ddots & \\ & & 0 & 1 \end{bmatrix}, \quad B_i = \begin{bmatrix} 1 & 0 & & \\ & \ddots & \ddots & \\ & & 1 & 0 \end{bmatrix}.$$

- *Blocks* (A_i, B_i) *of type* 5: $n_i = m_i - 1 \geq 1$ and $(A_i^\top, B_i^\top)$ is a block of type 4.

Mirwald shows in his dissertation that the following Kronecker normal form theorem also holds for the field $\mathbb{Z}_2$. A proof for this theorem in case of an infinite field of coefficients can be found in GANTMACHER (1959).

Theorem 12 *(Kronecker normal form theorem)*
Every pair $(A, B) \in M_{m,n}(\mathbb{Z}_2)^2$ *is equivalent to some pair in Kronecker normal form. Two pairs of matrices in Kronecker normal form are equivalent if and only if they coincide up to a permutation of their blocks, i.e. if they are formed by the same collection of blocks.*

Corollary 13 *Two pairs of alternate matrices* (A_1, B_1), $(A_2, B_2) \in M_{n,n}(\mathbb{Z}_2)^2$ *are congruent if and only if they are equivalent to the same Kronecker normal form.*

The rank $R(A, B)$ and the complexity $L(f_{\widetilde{A}}, f_{\widetilde{B}})$ depend on the occurrence of the following exceptional blocks in the Kronecker normal form of (A, B).

Definition The following five blocks are called *exceptional*:

- $\left(\begin{bmatrix} 0 & 1 \\ 1 & 0 \end{bmatrix}, \begin{bmatrix} 1 & 0 \\ 0 & 1 \end{bmatrix} \right)$ (a block of type 1),

- $\left(\begin{bmatrix} 0 & 0 \\ 1 & 0 \end{bmatrix}, \begin{bmatrix} 1 & 0 \\ 0 & 1 \end{bmatrix} \right)$ (a block of type 2),

- $\left(\begin{bmatrix} 1 & 0 \\ 0 & 1 \end{bmatrix}, \begin{bmatrix} 0 & 0 \\ 1 & 0 \end{bmatrix} \right)$ (a block of type 3),

- $([0 \ 1], [1 \ 0])$ (a block of type 4),

- $\left(\begin{bmatrix} 0 \\ 1 \end{bmatrix}, \begin{bmatrix} 1 \\ 0 \end{bmatrix} \right)$ (a block of type 5).

In the following let $(A, B) \in M_{r,s}(\mathbb{Z}_2)^2$ be an arbitrary pair of matrices and let t_i for $i = 1, \ldots, 5$ be the number of exceptional blocks of type i in the Kronecker normal form of (A, B). Recall that we associate with $A, B \in M_{r,s}(\mathbb{Z}_2)$ matrices $\tilde{A}, \tilde{B} \in M_{r+s,r+s}(\mathbb{Z}_2)$ with

$$\tilde{A} = \begin{bmatrix} 0 & A \\ 0 & 0 \end{bmatrix}, \quad \tilde{B} = \begin{bmatrix} 0 & B \\ 0 & 0 \end{bmatrix}.$$

Theorem 14 (MIRWALD (1991))
Let $(A, B) \in M_{r,s}(\mathbb{Z}_2)^2$ and let t_i for $i = 1, \ldots, 5$ be as above. If $t_1 = t_2 = t_3 = 0$ we have

(1) $\mathrm{R}(A, B) = \lceil \frac{1}{2} (\mathrm{R}(A) + \mathrm{R}(B) + \mathrm{R}(A \oplus B) + t_4 + t_5) \rceil$,

(2) $\mathrm{L}(f_{\tilde{A}}, f_{\tilde{B}}) = \mathrm{R}(A, B) = \lceil \frac{1}{2} \mathrm{R}(\tilde{A} \oplus \tilde{A}^{\mathsf{T}}, \ \tilde{B} \oplus \tilde{B}^{\mathsf{T}}) \rceil$.

In particular if $A, B \in M_{n,n}(\mathbb{Z}_2)$ and

$$\mathrm{R}(A) = \mathrm{R}(B) = \mathrm{R}(A \oplus B) = n$$

then we have $t_1 = t_2 = t_3 = t_4 = t_5 = 0$ and thus

$$\mathrm{R}(A, B) = \mathrm{L}(f_{\tilde{A}}, f_{\tilde{B}}) = \lceil 3n/2 \rceil.$$

The crucial assertion of Theorem 14 is the equality (1) and the lower bound $\mathrm{L}(f_{\tilde{A}}, f_{\tilde{B}}) \geq \mathrm{R}(A, B)$. The upper bound $\mathrm{L}(f_{\tilde{A}}, f_{\tilde{B}}) \leq \mathrm{R}(A, B)$ follows from Theorem 4. Next we show that equality (1) in Theorem 14 implies

$$\mathrm{R}(A, B) = \lceil \tfrac{1}{2} \mathrm{R}(\tilde{A} \oplus \tilde{A}^{\mathsf{T}}, \ \tilde{B} \oplus \tilde{B}^{\mathsf{T}}) \rceil. \qquad (2)$$

The Kronecker normal form of the pair $(\tilde{A} \oplus \tilde{A}^{\mathsf{T}}, \ \tilde{B} \oplus \tilde{B}^{\mathsf{T}})$ is the "union" of the Kronecker normal forms of the pairs (A, B) and $(A^{\mathsf{T}}, B^{\mathsf{T}})$, i.e. it consists of the union of the collection of blocks in the Kronecker normal forms of (A, B) and $(A^{\mathsf{T}}, B^{\mathsf{T}})$. The Kronecker normal form of $(A^{\mathsf{T}}, B^{\mathsf{T}})$ is obtained by transposing the blocks in the Kronecker normal form of (A, B). Hence the numbers $\tilde{t}_i$ for $i = 1, \ldots, 5$ for the pair $(\tilde{A} \oplus \tilde{A}^{\mathsf{T}}, \ \tilde{B} \oplus \tilde{B}^{\mathsf{T}})$ are obtained from the corresponding numbers t_i for (A, B) by

$$\tilde{t}_i = 2 t_i \quad \text{for } i = 1, 2, 3,$$

$$\tilde{t}_4 = \tilde{t}_5 = t_4 + t_5.$$

Therefore if $t_1 = t_2 = t_3 = 0$ we have $\tilde{t}_1 = \tilde{t}_2 = \tilde{t}_3 = 0$ and equation (1) in Theorem 14 also holds for $(\tilde{A} \oplus \tilde{A}^{\mathsf{T}},\ \tilde{B} \oplus \tilde{B}^{\mathsf{T}})$. Now the equality (2) follows from

$$\mathrm{R}(\tilde{A} \oplus \tilde{A}^{\mathsf{T}}) = 2\,\mathrm{R}(A),\quad \mathrm{R}(\tilde{B} \oplus \tilde{B}^{\mathsf{T}}) = 2\,\mathrm{R}(B),$$

$$\mathrm{R}(\tilde{A} \oplus \tilde{A}^{\mathsf{T}} \oplus \tilde{B} \oplus \tilde{B}^{\mathsf{T}}) = 2\,\mathrm{R}(A \oplus B),\quad \tilde{t}_4 + \tilde{t}_5 = 2(t_4 + t_5).$$

We finally consider the case that $t_1 + t_2 + t_3 \neq 0$. We call a block (A_i, B_i) *unsaturated* if $\mathrm{R}(A_i) + \mathrm{R}(B_i) + \mathrm{R}(A_i \oplus B_i)$ is odd. Note that all exceptional blocks are unsaturated. Let t_u denote the number of unsaturated blocks in the Kronecker normal form of (A, B), and let t denote the maximum of t_1, t_2, t_3.

Theorem 15 (MIRWALD (1991))
Let $(A, B) \in M_{r,s}(\mathbb{Z}_2)^2$ with t_i for $i = 1, \ldots, 5$, t, t_u as above. If $t_4 + t_5 + 2t \leq t_u$ then we have

(1) $\mathrm{R}(A, B) = \lceil \frac{1}{2}(\mathrm{R}(A) + \mathrm{R}(B) + \mathrm{R}(A \oplus B) + t_4 + t_5) \rceil$,

(2) $\mathrm{L}(f_{\tilde{A}}, f_{\tilde{B}}) = \mathrm{R}(A, B) = \lceil \frac{1}{2}\mathrm{R}(\tilde{A} \oplus \tilde{A}^{\mathsf{T}},\ \tilde{B} \oplus \tilde{B}^{\mathsf{T}}) \rceil$.

For the case that $t_4 + t_5 + 2t > t_u$ our lower and upper bounds do not coincide:

$$\lceil \tfrac{1}{2}(\mathrm{R}(A) + \mathrm{R}(B) + \mathrm{R}(A \oplus B) + t_4 + t_5) \rceil$$
$$\leq \mathrm{L}(f_{\tilde{A}}, f_{\tilde{B}}) \leq \mathrm{R}(A, B) \leq$$
$$\lceil \tfrac{1}{2}(\mathrm{R}(A) + \mathrm{R}(B) + \mathrm{R}(A \oplus B) + 2(t + t_4 + t_5) - t_u) \rceil.$$

For this case we can neither prove that $\mathrm{R}(A, B) = \mathrm{L}(f_{\tilde{A}}, f_{\tilde{B}})$ nor that $\mathrm{R}(A, B) = \lceil \frac{1}{2}\mathrm{R}(\tilde{A} \oplus \tilde{A}^{\mathsf{T}},\ \tilde{B} \oplus \tilde{B}^{\mathsf{T}}) \rceil$. The condition $t_4 + t_3 + 2t > t_u$ that characterizes the unsolved case is equivalent to the existence of a type $i \leq 3$, such that the number of exceptional blocks of type i is dominating in the sense that

$$t_i > \sum_{j \leq 3, j \neq i} t_j\ +\ \#\,\{\text{unsaturated, non–exceptional blocks}\}\,.$$

For the extreme case that the Kronecker normal form of (A, B) consists *only* of exceptional blocks MIRWALD (1991) shows that the above upper bound is tight. More precisely, in this extreme case we have

$$\mathrm{R}(A, B) = \lceil \tfrac{1}{2}(\mathrm{R}(A) + \mathrm{R}(B) + \mathrm{R}(A \oplus B) + 2(t + t_4 + t_5) - t_u) \rceil,$$
$$\mathrm{L}(f_{\tilde{A}}, f_{\tilde{B}}) = \mathrm{R}(A, B) = \lceil \tfrac{1}{2}(\mathrm{R}(\tilde{A} \oplus \tilde{A}^{\mathsf{T}},\ \tilde{B} \oplus \tilde{B}^{\mathsf{T}})) \rceil.$$

The proofs for the main results are rather technical. It is remarkable that the lower bound proofs are rather clean whereas the upper bounds require a tedious case analysis depending on the structure of the Kronecker normal form of the pair (A, B).

References

F. R. GANTMACHER: *The Theory of Matrices, Vol. 1 and 2.* Chelsea Publishing Company. New York 1959.

D. YU. GRIGORYEV: *Some New Bounds on Tensor Rank.* Lomi Preprints E-2-78. Leningrad 1978.

H. F. DE GROOTE: *Lectures on the Complexity of Bilinear Problems.* Lecture Notes in Computer Science, Vol. 245. Springer–Verlag. Berlin, Heidelberg, New York 1987.

J. HÅSTAD: *Tensor Rank is NP–complete.* Journal of Algorithms **11** (1990), pp. 644–654.

J. JA' JA': *Optimal evaluation of Pairs of Bilinear Forms.* SIAM J. Computing **8** (1979), pp. 443–462.

R. MIRWALD: *Dissertation,* Universität Frankfurt (1991).

R. MIRWALD, C. P. SCHNORR: *The Multiplicative Complexity of Quadratic Boolean Forms.* Proc. of 28th IEEE Symposium on Foundations of Computer Science, 1987, pp. 141–150. Final paper to appear in Theor. Comp. Sci..

V. STRASSEN: *Vermeidung von Divisionen.* Crelles Journal für reine und angew. Mathematik **264** (1973), pp. 184–202.

Some Problems Involving Razborov-Smolensky Polynomials

David A. Mix Barrington *

Abstract

Several recent results in circuit complexity theory have used a representation of Boolean functions by polynomials over finite fields. Our current inability to extend these results to superficially similar situations may be related to properties of these polynomials which do not extend to polynomials over general finite rings or finite abelian groups. Here we pose a number of conjectures on the behavior of such polynomials over rings and groups, and present some partial results toward proving them.

1. Introduction

1.1. Polynomials and Circuit Complexity

The representation of Boolean functions as polynomials over the finite field $Z_2 = \{0, 1\}$ dates back to early work in switching theory [Sh38]. A formal language L can be identified with the family of functions $f_i : Z_2^i \to Z_2$, where $f_i(x_1, \ldots, x_i) = 1$ iff $x_1 \ldots x_i \in L$. Each of these functions can be written as a polynomial in the variables $x_1, \ldots, x_n$. We can consider algebraic formulas or circuits with inputs $\{x_1, \ldots, x_n\}$ and the usual complexity measures of size and depth, and get a complexity theory with an algebraic character. Since the binary AND and XOR functions can be simulated in constant size and depth by the binary AND and OR functions and vice versa, this is essentially the usual complexity theory of Boolean circuits and formulas.

Similar notions have been considered for some time. Skyum and Valiant [SV81] introduced a complexity theory along these lines, but used the two-element Boolean algebra, rather than Z_2, as the base of their polynomials.

---

*Department of Computer Science, University of Massachusetts, Amherst MA 01003 U.S.A. The author was partially supported by NSF Computer and Computation Theory grant CCR-8714714.

Along with the conventional complexity measure of circuit size, they considered the "degree" of a circuit, based on the depth of AND gates in it. For example, their class pdC of polynomials with circuit size and degree each polynomial in n, is the class now known as (non-uniform) $LOGCFL$. This work emphasized non-uniform complexity classes and very simple reductions between languages.

In this paper we will use a different and more purely algebraic notion of "degree", which depends only on the polynomial and not on the circuit used to calculate it. Since the input variables are each idempotent, we will define the degree of a monomial to be the number of variables occurring in it. As usual, the degree of a polynomial will be the maximum degree of any monomial occurring in it with a nonzero coefficient.

Razborov [Ra87] used this algebraic view to develop a new lower bound technology for Boolean circuits. He noted that the functions in the class $AC^0[2]$ (those functions computable by constant-depth, poly-size, unbounded fan-in circuits of AND, OR, and XOR gates) share a certain algebraic property. Each such function can be approximated by a polynomial of relatively low degree (the function and the polynomial agree except on a small fraction of the inputs). He showed that the MAJORITY function cannot be so approximated, and thus confirmed the general belief that MAJORITY cannot be computed with such circuits.

Smolensky [Sm87] extended Razborov's method in two ways. First (along with Barrington [Ba86]) he developed an algebraic setting in which MOD-p, for a particular prime p, was a primitive operation instead of MOD-2. Given a field F of characteristic p, he defined the algebra of polynomials over F in variables $\{x_1, \ldots, x_n\}$ satisfying the identities $x_i^2 = x_i$. We will call this the Razborov-Smolensky ring of order n over F, $RS_n(F)$. (Later we will speak of the ring $RS_n(R)$ similarly defined with respect to any finite ring R, and even of the group $RS_n(G)$ defined with respect to any finite abelian group G.) A function from Z_2^n to F is represented by a unique polynomial in $RS_n(F)$, and (following Razborov) functions computable by constant-depth, poly-size, unbounded fan-in circuits of AND, OR, and MOD-p gates (the class $AC^0[p]$) can be approximated by low-degree polynomials in $RS_n(F)$ whenever F is of characteristic p. Razborov's method can then be extended [Ba86] to show that MAJORITY is not in $AC^0[p]$.

Smolensky's second and more important contribution was to show that the iterated multiplication function of F cannot be so approximated. This not only provides a simpler proof that MAJORITY is not in $AC^0[p]$, but shows that none of the functions MOD-q (with q prime to p) are in that class. (In retrospect, Smolensky's methods also provide a reasonably simple way to an earlier seminal result. This is the theorem of Furst-Saxe-Sipser [FSS84] and Ajtai [Aj83] that the functions MOD-q cannot be computed with constant-depth polynomial-size circuits of AND and OR gates (are not in the class AC^0)).

It is intriguing that these methods provide no way to place bounds on the computing power of such circuits with AND, OR, and MOD-q gates unless q is a prime or prime power. Indeed no non-trivial bounds on this class $AC^0[q]$ are known (it might be equal to NP) although no surprising functions are known to be in it. The union of $AC^0[q]$ for all integers q is called ACC^0 [MT89] (also earlier called ACC [BT88]). It is conjectured [Ba89] that MAJORITY is not in ACC^0 and thus that ACC^0 is a strict subset of NC^1 (circuits of depth $O(\log n)$ and fan-in two). Yao [Ya90] (see also [Al89], [BT91]) has recently shown that functions in ACC^0 can be computed by threshold circuits of depth 3 and quasipolynomial ($2^{(\log n)^{O(1)}}$) size, showing that either ACC^0 is small or such threshold circuits are very powerful.

1.2. The Programs-Over-Monoid Model

The same methods were subsequently used to obtain further lower bounds in circuit complexity, which are most easily stated using the language of programs over finite monoids [BT88]. A monoid is a set with an associative binary operation and an identity. In classical algebraic automata theory, the computation of a finite-state machine is viewed as an iterated multiplication in a particular finite monoid (the transformations of states of the machine, under the operation of composition). Every finite-state computable language has a particular minimal monoid (called the syntactic monoid) which must be contained within any finite-state machine which recognizes the language. A complexity theory of the regular languages has been developed, where one language is viewed as harder than another if its syntactic monoid contains that of the first. ("Contains" has a well-defined meaning — a monoid A contains B (or B divides A) if there is a monoid homomorphism from a submonoid of A onto B.)

Furthermore, there is a structure theory of finite monoids due originally to Krohn and Rhodes [KRT68]. Just as every integer is a product of primes, and every finite group (by the Jordan-Hölder Theorem) can be built up from simple groups in a certain way, every finite monoid can be constructed from particular building blocks using particular operations. The building blocks are the simple groups (Z_p for prime p and various non-abelian groups) and one other monoid which is not a group. The operations are division and the *wreath product* — for details see [BT88] and [BST90].

To get natural subsets of the set of finite monoids (and hence natural sub-classes of the regular languages), we can restrict which building blocks may be used in the wreath product operation. In this way we can get the classes of solvable groups (only groups of the form Z_p), all groups (only simple groups), aperiodic monoids (only the non-group component), solvable monoids (the non-group component and Z_p's), or all monoids (all components). For each such set of monoids we can look at the class of regular languages whose syntactic monoids lie within that set.

What does all this have to do with circuit complexity? It turns out that the structure of these classes of monoids and regular languages bears a close relationship to that of various circuit complexity classes [BT88]. Given a monoid M and a set $\{x_1, \ldots, x_n\}$ of Boolean inputs, define an *instruction* as the name of an input and a map from Z_2 to M (i.e., two elements of M). The *yield* of an instruction (i, a, b), say, is a if $x_i = 0$ and b if $x_i = 1$. A *program* over M is a sequence of instructions, and its yield is the ordered product of the elements of M yielded by each instruction. A program thus defines a map from Z_2^n into M, or a Boolean function on n variables if we divide the elements of M into "accepting" and "rejecting".

Barrington [Ba89] showed that poly-length programs over the non-abelian simple group A_5 (which correspond to poly-length width-5 branching programs obeying certain restrictions) can be constructed to recognize any language in the circuit complexity class NC^1. NC^1 is easily seen to be powerful enough to simulate polynomial-length programs over any finite monoid, so it is equal to the class of languages recognized by such programs. The circuit complexity classes AC^0, $AC^0[p]$, and ACC^0 discussed above (each of which is a subclass of NC^1), are exactly the classes of languages recognized by poly-length programs over aperiodic monoids, p-monoids, and solvable monoids respectively [BT88]. (A p-monoid is one built up from the non-group component and Z_p in the Krohn-Rhodes framework, where p is prime.)

The connection between circuits and monoids is somewhat understood — the operations of modular and threshold counting occur in the same places in both settings, and the placing of one operation above another in a circuit corresponds to the wreath product operations. The classes of regular languages can also be thought of as very uniform versions of the corresponding circuit complexity classes, especially when both are considered as expressibility classes in the sense of Immerman [BIS90]. Other complexity classes defined by programs have also been shown to be of interest. Szegedy [Sz90] has shown that the languages recognized by programs over abelian monoids are exactly the languages of constant symmetric communication complexity. Bedard, Lemieux, and McKenzie [BLM90] have defined an extension of the model to programs over finite *groupoids* (binary algebras, not necessarily associative) and shown that poly-length programs over groupoids recognize exactly the class $LOGCFL$.

1.3. Polynomials and Programs over Groups

Barrington, Straubing and Thérien [BST90] have begun the study of the computational power of programs over finite groups. It is known that poly-length programs over any non-solvable group compute exactly NC^1, and that poly-length programs over any solvable group compute only languages in ACC^0. Since we believe that $ACC^0 \neq NC^1$, we believe that programs over solvable monoids are relatively limited in power. The power of solvable monoids can be studied in two stages: the power of solvable groups and the possible interactions between the group and non-group components of a monoid.

In particular, it is conjectured [BST90] that no poly-length program over a solvable group can compute the AND of all n variables. Given our polynomial language, this is fairly easily seen to be true for p-groups. A program over a p-group, of whatever length, can be simulated by a polynomial of constant degree over Z_p. If, for some f of constant degree, $f = a$ iff $x_1 x_2 \ldots x_n = 1$, then the polynomial $1 - (f - a)^{p-1}$ is also of constant degree and must be exactly $x_1 x_2 \ldots x_n$, which is impossible unless n is a constant.

Even moving to nilpotent groups (direct products of p_i-groups, in general involving more than one p_i), things are not so easy. In the case of programs over Z_6, for example, we can convert such a program into one polynomial over Z_2 and one over Z_3, each of constant degree. We know about the sets defined by each of these two polynomials, but how do we rule out the possibility that two of these sets intersect in the singleton $(1, 1, \ldots, 1)$? A Ramsey argument [BST90] shows that this is impossible for "sufficiently large", but still constant, n.

Over groups which are solvable but not nilpotent, it is possible for a program of exponential length to compute the AND of all the variables. For some of these groups, methods have been devised to show that exponential length is necessary, but in general this remains open. The known subcases are groups which divide the wreath product of a p-group and an abelian group [BST90], and dihedral groups [St91]. A plausible but unproven conjecture about the ring $RS_n(Z_p)$ would extend the exponential lower bound to groups dividing the wreath product of a p-group and a nilpotent group [BST90].

These constructions, like Smolensky's, use the ring $RS_n(F)$ for a field F of characteristic p having elements of the desired multiplicative order. Rather than degree, the complexity measure used on polynomials is the number of terms needed to form them by addition, in two different bases of $RS_n(F)$.

The Razborov-Smolensky ring of polynomials can be defined over an arbitrary ring as well as over a field. The set of such polynomials over an arbitrary abelian group still forms a group under polynomial addition, though it no longer has a ring structure. Many properties, such as the unique representation of any function from Z_2^n to R by a polynomial, still hold, but some do not. One which does not is the one we used above to convert from one way to represent a set by a polynomial to another:

Definition 1.1 *A polynomial f in $RS_n(R)$ weakly represents a set $A \subseteq Z_2^n$ if for some $a \in R$, $f(\mathbf{x}) = a$ iff $\mathbf{x} \in A$. A polynomial f strongly represents a set A if $f(\mathbf{x}) = 1$ for $\mathbf{x} \in A$ and $f(\mathbf{x}) = 0$ otherwise.*

Fact 1.2 *If f weakly represents A with value a over a finite field F, then $1 - (f - a)^{|F|-1}$ strongly represents A and has degree at most a constant multiple of that of f.* □

We know that this exact property can fail for polynomials over a ring which is not a field. Consider $R = Z_6$, and let $f = x_1 + x_2 + \ldots + x_n$. This linear polynomial weakly represents the set $A = \{\mathbf{x} : x_1 + x_2 + \ldots x_n = 0 \bmod 6\}$. But the unique polynomial strongly representing A over Z_6 can be calculated (using methods in [BST90] or [Sm87]) — it is congruent mod 3 to the MOD-2 function and congruent mod 2 to the MOD-3 function. It has degree n, the maximal possible degree.

The inability to do this conversion occurs as a roadblock constantly in the algebraic approach to circuit lower bounds. In this paper we would like to raise the possibility that some relationship between weak and strong representation might hold for other reasons, and that this might allow these algebraic methods to be extended.

2.   The Small Image-Set Conjecture

Let us first consider attempting to represent a small set with a polynomial of as small a degree as possible. (We will define the *size* of a subset $A \subseteq Z_2^n$ to be the fraction of the entire set it contains, i.e., $|A|/2^n$.) For strong representation degree n is needed because the unique polynomial representing that function has degree n (and an exponential number of terms). With degree d we can strongly represent a set of size 2^{-d}, but not one any smaller (we'll prove this later). We can do somewhat better if we are willing to settle for weak representation. For example, in Z_m we can have $m - 1$ terms of degree $\lceil n/(m-1) \rceil$, each giving one if its variables are all one and zero otherwise. The sum of these terms gives $m - 1$ iff all n variables are one. A reasonable first conjecture says that this construction is the best possible.

Conjecture 2.1 *(Small Image-Set Conjecture, Weak Form) Any nonempty set weakly represented by a polynomial of degree d over R has size at least $2^{-d(|R|-1)}$.*

Furthermore, something stronger seems to be true. If we consider the number of values taken on by the polynomial, we have a continuum between two (strong representation) and $|R|$ (weak representation). Using examples similar to those above and conjecturing that they are best possible, we are led to:

Conjecture 2.2 *(Small Image-Set Conjecture) Any nonempty set weakly represented by a polynomial of degree d which takes on r values has size at least $2^{-d(r-1)}$.*

Note that neither of these conjectures makes any reference to the multiplicative structure of $RS_n(R)$ — in fact they are conjectures about the group $RS_n(G)$, where G is the additive group of R. (The use of the letter R below will not necessarily imply the existence of a ring structure.)

Unfortunately, both these conjectures fail in the case $d = 3$ and $R = Z_6$. If we let s_i denote the sum of all terms of degree i in 27 variables over Z_6, the polynomial $s_3 + 5s_2 + 3s_1$ weakly represents the singleton set containing the all-zero element of Z_2^{27} (this example was discovered by David Applegate, Jim Aspnes, and Steven Rudich). This set has size 2^{-27}, smaller than the 2^{-15} predicted by Conjecture 2.1.

Barrington, Beigel, and Rudich [BBR91] have shown that this is an example of a general phenomenon for *symmetric* polynomials over an abelian group (where the monomials of a given degree must all have the same coefficient, with an absent monomial considered to have zero coefficient). Furthermore, this behavior is the best that symmetric polynomials can do, so that they satisfy a variant of Conjecture 2.1. To state their result, let $\rho(g)$ be the number of distinct primes dividing the order of an abelian group G, so that $\rho(Z_6) = 2$.

Fact 2.3 *[BBR91] The minimum size of a set weakly represented by a symmetric polynomial of degree d over G is $2^{-\Theta(d^{\rho(G)})}$.* $\square$

Conjecture 2.4 *The minimum size of a set weakly represented by* any *polynomial of degree d over G is $2^{-\Theta(d^{\rho(G)})}$.*

This question remains open. Here we will continue with the progress made on the original conjectures before this work, commenting on the later results as necessary.

Observation 2.5 *The SIS Conjecture is trivially true if $r = 1$, $d = 0$, or if $n \leq d(r - 1)$.* $\square$

To prove the SIS conjecture in some more interesting special cases, let us consider the following construction. Let f be a polynomial of degree at most d on n variables, taking on at most r values. Write f as $g + hx_n$ so that g and h are independent of x_n, and let $g' = g + h$. Note that h is of degree at most $d - 1$. The values taken on by f are exactly that of g (if $x_n = 0$) or of g' (if $x_n = 1$), so each of these functions is at most r-valued. If both take on the same r values, and the SIS Conjecture holds for smaller values of n, then it continues to hold for this n. This is because a set weakly represented by f in this case is the union of two subsets of Z_2^{n-1}, each weakly represented by a polynomial which is degree d and r-valued. But it is entirely possible for g, say, to take on only $r - 1$ values, so that some value is taken on only with $x_n = 1$. In this case the size of the subset of Z_2^n on which f takes that value would be half of that of the subset of Z_2^{n-1} where g' takes it on. In two special cases, however, we can see how to carry out an inductive proof of the Conjecture.

Fact 2.6 *The SIS Conjecture holds in the case $r = 2$.*

Proof : Let f, g, h, g' be as above. The argument above suffices if g and g' each take on both the values taken on by f. But because $r = 2$, if either g or g' fails to do so, it must be a constant. But then the other differs from a constant by h and is thus of degree at most $d - 1$. If we also induct on d, we can assume that a nonempty subset of Z_2^{n-1} weakly represented by g or g' has size at least $2^{-(d-1)}$, so that the same set viewed as a subset of 2^n has size at least 2^{-d}. $\square$

Fact 2.7 *The SIS Conjecture holds in the case $d = 1$.*

Proof : Again we use the terminology above. If $d = 1$, then h must be a constant. This means that g and g' take on the same number of values. If this number is r, then each takes on all the values of f and the argument above suffices. Otherwise each is at most $(r - 1)$-valued, and weakly represents sets of size only at least $2^{-(r-2)}$ in Z_2^{n-1}, which are of size $2^{-(r-1)}$ in Z_2^n. $\quad\square$

Fact 2.8 *The SIS Conjecture holds when R is a field.*

Proof : Here we can show that a set weakly represented by an r-valued f of degree d is also weakly represented by a two-valued polynomial of degree $d(r - 1)$. (Over a field, this is the same as being strongly represented by a polynomial of that degree, because we can divide by any constant.) The desired result will then follow from Fact 2.6 above.

Suppose $f = a$ on the set in question, and on other points takes on some value in $\{b_1, \ldots, b_{r-1}\}$. Consider the product over all i of $f - b_i$. It equals zero off the set in question, and equals a particular nonzero constant on it — the product over all i of $a - b_i$. $\quad\square$

Fact 2.9 *Conjecture 2.1 (the weaker form of the SIS Conjecture) holds when $R = Z_{p^k}$ for some prime p.*

Proof : A trick of Chandra, Stockmeyer, and Vishkin [CSV84], relating the MOD-p^k operation to the MOD-p operation, can be adapted to this setting. The sum modulo p^k of a set of Boolean terms is zero iff the modulo p sum of the terms is zero and the modulo p^{k-1} sum of all the products of p-tuples of the terms is also zero. Using this fact, we can take a degree d polynomial f over $R = Z_{p^k}$ and create a two-valued degree $d(p^k - 1)$ polynomial f' over Z_p such that $f = 0$ iff $f' = 0$. $\quad\square$

This means that the full SIS conjecture holds for the case $r = p^k$, but it is not known to hold for general r. In the case $r = 3$, if the range of f is $\{0, a, b\}$, the degree $2d$ polynomial $(f - a)(f - b)$ is two-valued and tests $f = 0$, unless $ab = 0$. The latter is possible only if p divides both a and b. But in that case f can be divided by p (a polynomial takes on all its values in a subgroup iff every term of it does), without changing its degree. In fact, for $r \leq 3$ and $r = p^k$, an r-valued polynomial of degree d over Z_{p^k} can be simulated by a two-valued degree $d(r - 1)$ polynomial over Z_p. It is not know whether this holds for all r (it is interesting that so far whether it does appears to be independent of k). Note also that the cases shown here suffice to prove the full SIS conjecture for $R = Z_4$.

This makes $R = Z_6$ the smallest ring for which the minimum size of a weakly representable set is unknown. We don't know whether a quadratic polynomial over Z_6 can weakly represent a singleton set with $n = 11$ (it can for $n = 10$). We don't even know whether a three-valued quadratic polynomial can weakly represent a singleton with $n = 5$ ($x_1x_2+x_3x_4$ suffices for $n = 4$). In the special case of quadratic polynomials, no symmetric polynomial can weakly represent a singleton with $n > 9$, so the new results of [BBR91] do not affect this case. But as described above, they do make a difference at $d = 3$, where a singleton can be weakly represented with $n = 27$, as opposed to the conjectured $n = 15$. No symmetric polynomial can do better, but whether an asymmetric one can is unknown.

The quadratic case has an alternate representation as a graph problem. Label both the vertices and edges of an undirected graph with elements of R, and consider the sums of the vertex and edge labels in a clique. What is the largest graph for which every nonempty clique has a nonzero sum? Heath and Pemmaraju [personal communication] have looked at such graphs (for cyclic R) and defined "tight" graphs to be those where (1) every nonempty clique sum is nonzero and (2) the number of distinct clique sums is minimal over all graphs of that size satisfying (1). They conjecture that tight graphs with r distinct nonzero clique sums have $2r$ vertices. They speculate that tight graphs might necessarily possess some structure which would allow an inductive proof on r. Their conjecture is independent of the modulus, suggesting that it may have a combinatorial rather than an algebraic proof. A proof of the SIS conjecture in this quadratic case might still lead to progress on the general question, though the original SIS conjecture does not hold.

3.   The Intersection Conjecture

One way to examine the sets of input values which can be defined by specific types of polynomials is to see how they intersect. For example, suppose that we have k equations $f_i = 0$, each of degree d. The SIS conjecture would give us limits on how small the locus of each equations could be, but would place no limits on the size of the intersection. Over a field, as we shall see below, we can use the multiplication to construct a single equation $f^\star = 0$ which is true iff $f_i = 0$ for all i. (One way to do this, which is not the most efficient, would be to raise each f_i to the $|F| - 1$'th power to get an f_i' which is always zero or one, and then let F be the product over all i of $1 - f_i'$.) To what extent is this a general phenomenon for rings or abelian groups?

We know that if R is not a field, then two sets which are each strongly represented by a polynomial of low degree can intersect in a set whose strong representation has high degree. Let $R = Z_6$ and consider two sets of inputs: those summing to zero mod 2 and those summing to zero mod 3. The strong representations of these sets over Z_6 are linear and quadratic, respectively, but that of their intersection is degree n. In this case the intersection has a weak representation of low degree. Does this always happen?

Conjecture 3.1 *(Intersection Conjecture, Strong Form) If $\{f_i\}$ is a set of k polynomials of degree at most d over a finite group G, then there is a polynomial $f^\star$ of degree at most kd such that $f^\star = 0$ iff $f_i = 0$ for all i.*

Even if this is false, the following weaker statement might still be true:

Conjecture 3.2 *(Intersection Conjecture, Weak Form) Let $\{f_i\}$ be as above, and let S be the set of inputs $\mathbf{x}$ such that $f_i(\mathbf{x}) = 0$ for all i. If S is nonempty, then there is a polynomial $f^\star$ of degree at most kd such that the set $\{\mathbf{x} : f^\star(\mathbf{x}) = 0\}$ is also nonempty but is no bigger than S.*

Proposition 3.3 *Conjecture 3.1 holds when G is a finite field F.*

Proof :   We will show that there is a polynomial g_k of degree k over F, in k variables which range over F (not just Z_2), such that $g_k(y_1, \ldots, y_k) = 0$ iff $y_i = 0$ for all i. Then we can let $f^\star$ be $g_k(f_1, \ldots, f_k)$, a polynomial of degree kd in the input variables with the desired property.

For any F and k there is another finite field E which is a vector space of dimension k over F. Fix a basis $\{w_1, \ldots, w_k\}$ of E, so that each $a \in E$ can be written as $a = y_1 w_1 + \ldots + y_k w_k$ and thus identified with the k-tuple $\{y_1, \ldots, y_k\}$ of elements of F. The operation of multiplication by a is a linear transformation of this vector space represented by a matrix M_a. The determinant of M_a is an element of F, which is zero iff a is the zero element of E. M_a is given in terms of the y_i as $y_1 M_{w_1} + \ldots + y_k M_{w_k}$, so that M_a is a matrix each of whose entries is a linear polynomial in $\{y_1, \ldots, y_k\}$ over F. The determinant of this matrix is a polynomial of degree k over F and is zero iff all its inputs are zero. $\square$

Note that although we used the multiplicative structure of F in the course of this proof, the statement of Conjecture 3.1 makes no reference to this structure. Thus we know that the conjecture holds for any abelian group which can be given a field structure, i.e., any product Z_p^e for p prime.

Observation 3.4 *If Conjecture 3.1 holds for two abelian groups G and H, it holds for their direct product.*

Proof :   Write each f_i as $\langle g_i, h_i \rangle$, with g_i over G and h_i over h. Each g_i has degree at most k, so there is a polynomial $g^\star$ over G of degree kd such that $g^\star = 0$ iff $g_i = 0$ for all i. Similarly there is an $h^\star$ over H of degree kd with $h^\star = 0$ iff $h_i = 0$ for all i. If $f^\star = \langle g^\star, h^\star \rangle$, then $f^\star = 0$ iff $g_i = 0$ and $h_i = 0$ for all i, which is true iff $f_i = 0$ for all i.   □

Corollary 3.5 *Conjecture 3.1 is true for any abelian group which is a direct product of cyclic groups of prime order, or equivalently any group whose exponent (least common multiple of the orders of the elements) is square-free.*
□

Proposition 3.6 *Conjecture 3.2 is true for groups of the form Z_{p^e}.*

Proof :   By the proof of Fact 2.9 in the previous section, a set weakly representable in degree d over Z_{p^e} is weakly representable in degree $d(p^e - 1)$ over Z_p. By Proposition 3.3 above, the intersection of k such sets is also weakly representable in degree $kd(p^e - 1)$ over Z_p. If this intersection is nonempty, by Fact 2.8, its size is at most $2^{-kd(p^e-1)}$. But over Z_{p^e} it is easy to construct a polynomial of degree kd which weakly represents a set of this size — take the sum of $p^e - 1$ independent monomials of degree kd. (If there are not enough variables available to do this, it is still possible to weakly represent a singleton set, which has the minimum possible size of any nonempty set.)
□

We do not know that the groups which satisfy Conjecture 3.2 are closed under direct product, so we still do not know whether it holds for Z_{12}, for example. The argument used for Observation 3.4 does not work.

The intuition here is that a polynomial of a certain degree has only so much power to isolate a small fraction of the input space. Our conjectures say in effect that the polynomial can do no better at isolating a small fraction of a subset of that space defined by another polynomial.

4.   Making Change in an Abelian Group

A family of k equations over a finite abelian group G can equally well be thought of as a single equation over the direct product G^k. If the original equations are all of degree d, then so is the composite equation. This opens the possibility of applying the SIS Conjecture to the study of systems of equations. For example, we know that for $d = 1$ and for any G, a family of k linear equations over G gives a single equation over G^k. A set weakly represented by this equation must be of size at most $2^{-(|G|^k-1)}$, because the SIS Conjecture holds with $d = 1$. But if our Intersection Conjectures are correct, we can do better. The set of inputs satisfying each component has size at least $2^{-(|G|-1)}$, so the intersection of these sets would be size $2^{-k(|G|-1)}$. Perhaps a stronger form of the SIS Conjecture holds for many-fold direct products? The following problem is an attempt to investigate this. We restrict ourselves to the special case of linear polynomials, and as before we ignore any multiplicative structure which G might have and consider it as an abelian group.

Definition 4.1 *The* Boolean span *of a multiset of elements of G is the set of all sums of submultisets of it. The* proper span *of a multiset of nonzero elements is the set of all sums of nonempty submultisets.*

Problem 4.2 *What conditions on a multiset from G force its Boolean span to be all of G?*

We call this problem "change-making" because it is a variation of a familiar problem: How many coins of a specific type are needed to guarantee having change available for any amount? Rather than have the exact number of pence available for our purchase, we might be satisfied to have the correct number modulo 100, so we would be able to transfer an integral number of pounds. Thus we might want the Boolean span of our coins in Z_{100} to be all of Z_{100}.

A similar problem has been studied in the number theory literature – how large must a multiset from G be before its proper span includes 0? This number is at most one greater than the answer to our problem. Olson [O169, O169a] answered this question for p-groups and for the direct product of two cyclic groups where the order of one divides the order of the other. Baker and Schmidt [BS80] gave an upper bound on this number for general groups which is within a logarithmic factor of the known lower bound in many important cases. We will present our independent treatment of the question here and discuss the consequences of their results as necessary and in our conclusion.

Lemma 4.3 *Any multiset of $p - 1$ nonzero elements from Z_p has Boolean span Z_p.*

Proof : By induction on k, we show that any set of k coins spans a set of size at least $k + 1$ in Z_p, if $k < p$. The $k = 0$ case is trivial. Given k coins spanning a set S and a new coin a, we can make all of S and also the set $S + a = \{s + a : s \in S\}$. $S + a$ has the same size as S and must be different from S (S could be invariant under translation by a nonzero a only if it were a subgroup). So the new Boolean span is strictly larger than S, and has size at least $k + 2$ if S has size at least $k + 1$. □

Corollary 4.4 *Any multiset of size at least $p - 1$ of nonzero elements of Z_p has a submultiset of size $p - 1$ whose Boolean span is Z_p.* □

We can apply this Corollary to get a lower bound on the size of a nonempty set weakly represented by a linear polynomial. If the linear polynomial is given by $\sum_{i=1}^{n} c_i x_i$, choose a subset of the c_i's spanning Z_p. The terms with these coefficients (with the other variables zero) give all possible values and take on each value on a set of size at least $2^{-(p-1)}$. Adding in the other variables one by one cannot give rise to a smaller set. Of course this is just the $d = 1$ case of the SIS conjecture. But we will see that improved solutions to the change-making problem will give better bounds in the case of more general groups.

Applying this same argument to other cyclic groups, we get a more complicated result, because these groups might have nontrivial subgroups:

Lemma 4.5 *Any multiset of $m - 1$ elements of Z_m has a nonempty submultiset whose Boolean span is a subgroup of Z_m.*

Proof : A zero element spans the trivial subgroup. The process from the Lemma above must terminate with S a proper subgroup or continue until the span is all of Z_m. □

This means that having $m - 1$ "coins" no longer guarantees having change for all amounts, but any larger set must contain redundant coins.

Corollary 4.6 *Any multiset M of size at least $m - 1$ elements of Z_m has a submultiset of size at most $m - 1$ with the same Boolean span as M.*

Proof : Assume by induction that the Corollary is true for all Z_k with $k < m$. Using the Lemma, choose a submultiset N of M spanning a subgroup H, such that no proper submultiset of N spans a subgroup. Discard from M

any other elements of H. If $H = Z_m$ then N must have only $m - 1$ elements as otherwise we could apply the Lemma again to a proper submultiset of it. Otherwise H is isomorphic to Z_k for some $k < m$ and has at most $k - 1$ elements. Now discard from M all but one element in each coset of H. of the subgroup. This has no effect on the span as the difference can be made up by the elements generating the subgroup. We are left with at most $(k - 1) + (m/k - 1) < m - 1$ elements in M.   □

As above, we can use the Corollary to show that any nonempty set weakly represented by a linear polynomial over Z_m has size at least $2^{-(m-1)}$. This is also a known special case of the SIS Conjecture. But for groups which are not cyclic, a stronger change-making result appears to hold, which could be used to get better lower bounds on the size. The key parameter of the group is not the order but, roughly, the sum of the orders of the ring's components.

Definition 4.7 *Let R be a finite abelian group. Let e be the maximum order of any element of R. The reader may verify that R can be written as a direct product $Z_e \times R'$. Define a function on abelian groups recursively $f(Z_1) = 0$ and $f(R) = (e - 1) + f(R')$. We will call this function the* capacity *of the group.*

Fact 4.8 *For any R, there is a multiset of $f(R)$ elements such that (1) its Boolean span is R, (2) its proper span is $R \setminus \{0\}$.*

Proof :   To get (1) and (2), write R as $Z_e \times R'$ and let the multiset be $e - 1$ copies of $(1, 0_{R'})$ together with $(0, x)$ for each x in the set for R'.   □

This means that a linear polynomial over R can weakly represent a set of size $2^{-f(R)}$. We believe that no polynomial can do better.

Conjecture 4.9 *No multiset of more than $f(R)$ elements from R satisfies conditions (1) and (2) in the Fact above. That is, no linear polynomial over R can weakly represent the function strongly represented by $ax_1x_2 \ldots x_{f(R)}$ (where $a \in G$).*

Olson proves that if G is a p-group [Ol69] or the direct product of two cyclic groups the order of one of which divides the order of the other [Ol69a], then any multiset of size $f(G) + 1$ has 0 in its proper span. Thus Conjecture 4.9 is true for these groups.

A natural way to try to prove Conjecture 4.9 in general would be to use induction on subgroups. We could duplicate the argument for cyclic groups above if we knew that every multiset of $f(R)$ elements has a subset spanning a subgroup H. Unfortunately, this is not true: Consider the multiset from $Z_2 \times Z_4$, listed as $\{(0,1),(0,1),(1,1),(1,1)\}$. No subset spans a subgroup (the sum $(1,0)$ cannot be formed). Similar examples exist in many other rings. A related property, however, is true in so far as we have been able to check and suffices for our purposes:

Conjecture 4.10 *(The Change Conjecture) Any multiset of $f(R) + 1$ elements from R has a submultiset of size at most $f(R)$ whose Boolean span is a subgroup.*

Corollary 4.11 *(assuming Conjecture 4.10) Any multiset M of more than $f(R)$ elements from R has a submultiset of size at most $f(R)$ whose Boolean span is the same as that of M.*

Proof :   Assume by induction that the Corollary is true for all groups smaller than R. Apply the Conjecture to M to get a submultiset N of size at most $f(R)$ whose span is a subgroup H. If $H = R$ we are done. Otherwise replace N by a submultiset of size at most $f(H)$ which still spans H, by the inductive hypothesis. Discard any other elements of H in M. Now consider the remaining elements of M as elements of R/H (as their effect on the span of M depends only on which coset they are in because any element of H can be formed). Using the inductive hypothesis again, find a submultiset of these of size at most $f(R/H)$ with the same span and discard the others. We are left with at most $f(H) + f(R/H) \leq f(R)$ elements.   □

¿From this Corollary we get an improved bound for weak representation just as before:

Corollary 4.12 *(assuming Conjecture 4.10) Any nonempty set weakly represented by a linear polynomial over R has size at least $2^{-f(R)}$.*   □

These last conjectures imply the weaker form of the Intersection Conjecture with $d = 1$, because k equations over R are equivalent to one equation over R^k, and $f(R^k) = k f(R) \leq k(|R| - 1)$.

5.   Consequences

Our primary goal in this study is to extend our understanding of the computational power of Razborov-Smolensky polynomials over general finite rings and groups. We believe that resolving the conjectures here will require techniques which should be of general utility in the study of lower bounds for the related computational models of programs over groups and Boolean circuits. But there are some direct consequences which would ensue if certain of our conjectures were to be proven.

Though the original SIS conjecture is false, we have a revised version that includes the number of distinct primes dividing the order of the group, essentially saying that the symmetric functions of [BBR91] are the best possible. A proof of this would show that programs over nilpotent groups cannot calculate functions whose polynomials have more than constant degree. In particular they cannot compute the AND function on more than a constant number of variables. Both of these facts are currently known, but the proof of them [BST90] involves a Ramsey argument. Because of this, the "constant" bound on the number of variables for a given nilpotent group is very large. The new proof would reduce this bound to a polynomial in the order of the group (whose degree is the number of primes dividing the order).

It is possible that similar techniques would allow the extension of the exponential lower bound for program length to groups which are the wreath products of nilpotent groups with abelian groups. This is because any such group divides a direct product of groups of the kind we can currently handle. But since other parameters than degree are used in the existing lower bound proofs, the new results would have to deal with them.

Our conjectures, even to the extent that they are already proven, have minor consequences in circuit complexity. Consider a circuit of MOD-6 gates computing the AND function (a MOD-6 gate, which has unbounded fan-in, outputs one iff the MOD-6 sum of its inputs is nonzero, and otherwise outputs zero). It is conjectured that such a circuit would require an exponential number of gates (this is implied by the conjecture that programs over solvable groups require exponential length to compute AND). Can we prove any kind of lower bound on the number of gates?

Prior to this work, the best known result along these lines was that of Smolensky [Sm90], who showed that $\Omega(\log n)$ gates are required. We can duplicate this result, and also show that $\Omega(\log n)$ gates must occur on the level nearest the inputs, as follows.

If k is the number of gates on this level, then the level as a whole computes a function from Z_2^n to Z_6^k (before converting the output of this function to an element of Z_2^k). This function can be represented by a linear polynomial in the n Boolean input variables over the ring Z_6^k. By the $d = 1$ case of the SIS Conjecture, which we have proven, a nonempty set weakly represented by this linear function has size at least $2^{-(6^k-1)}$. For the circuit to compute the AND function, the first row must behave differently on the input $(1, 1, \ldots, 1)$, so that the polynomial for the first row must weakly represent a singleton set and thus k must be $\Omega(\log n)$.

The possibility of a better bound originally motivated much of the work here. Since the Intersection Conjecture holds for Z_6, we know that a linear polynomial over Z_6^k is equivalent to a polynomial of degree k over Z_6. Thus, the (false) SIS conjecture for $R = Z_6$ would tell us that $k = \Omega(n)$, and the (unproven) revised one would give $k = \Omega(\sqrt{n})$.

The (unproven) Change Conjecture for $R = Z_6^k$ would also give a linear lower bound, by the argument presented at the end of the last section. But although we do not know that a multiset of $f(Z_6^k) = 5k$ elements of Z_6^k must satisfy the conclusion of the Change Conjecture, it is not hard to show that a set of size $\Omega(k^2)$ must do so. Again by the argument in the last section, this yields an $\Omega(\sqrt{n})$ lower bound on the number of gates in a MOD-6 circuit for AND. Baker and Schmidt [BS80] improve this parameter to $\Omega(k \log k)$, yielding an $\Omega(n \log n)$ bound on the circuit size.

The best current result, however, uses other methods. Thérien [Th91] has recently proved an $\Omega(n)$ lower bound on the number of gates required to compute the AND function with a depth-2 circuit of MOD-q gates, for any q not a prime power. His short proof relies on the machinery developed in [BST90].

6.   Acknowledgements

Much of this paper represents joint work with Denis Thérien, who suggested a number of these problems. I would like to thank him, the other participants in the 1989 and 1990 McGill Invitational Workshops, the participants in the 1990 Durham Symposium (especially Imre Leader), my colleagues in the COINS Theory Group, David Applegate, Jim Aspnes, Richard Beigel, Johan Håstad, Lenny Heath, Sriram Pemmaraju, and Steven Rudich for various helpful discussions. The two anonymous referees made a number of helpful suggestions and corrections which have improved the presentation. In addition, one of them provided the proofs of Proposition 3.3 and Observation 3.4, improving the results from the earlier version.

References

[Aj83] Ajtai M., Σ_1^1 formulae on finite structures. Annals of Pure and Applied Logic 24 (1983), 1-48.

[Al89] Allender E., *A note on the power of threshold circuits*. Proceedings of the *30th IEEE FOCS Symposium (1989), 580-585*.

[BBR91] Barrington D.A.M., Beigel R., Rudich S., *Representing Boolean functions as polynomials modulo composite numbers. Submitted to 1992 FOCS, also COINS Technical Report 91-82 (Nov. 1991), University of Massachusetts*.

[BS80] Baker R.C., Schmidt W.M., *Diophantine problems in variables restricted to the values 0 and 1*. J. of Number Theory 12 (1980), 460-486.

[Ba86] Barrington D.A., *A note on a theorem of Razborov*. COINS Technical Report 87-93 (July 1986), University of Massachusetts.

[Ba89] Barrington D.A., *Bounded-width polynomial-size branching programs recognize exactly those languages in NC^1*. J. Comp. Syst. Sci. 38:1 (Feb. 1989), 150-164.

[BIS90] Barrington D.A.M., Immerman N., Straubing H., *On uniformity within NC^1*. J. Comp. Syst. Sci. 41:3 (Dec. 1990), 274-306.

[BST90] Barrington D.A.M., Straubing H., Thérien D., *Non-uniform automata over groups*. Information and Computation 89:2 (Dec. 1990), 109-132.

[BT88] Barrington D.A.M., Thérien D., *Finite monoids and the fine structure of NC^1*. J. ACM (Oct. 1988), 941-952.

[BLM90] Bédard F., Lemieux F., McKenzie P., *Extensions to Barrington's M-program model. Structure in Complexity Theory: Fifth Annual Conference (1990), 200-209*.

[BT91] Beigel R., Tarui J., *On ACC*. Proceedings of the 32nd IEEE FOCS Symposium (1991), 783-792.

[CSV84] Chandra A.K., Stockmeyer L.J., Vishkin U., *Constant depth reducibility. SIAM J. Comp. 13:2 (1984), 423-439*.

[FSS84] Furst M., Saxe J.B., Sipser M., *Parity, circuits, and the polynomial-time hierarchy. Math. Syst. Theory 17 (1984), 13-27*.

[KRT68] Krohn K.B., Rhodes J., Tilson B., in Arbib M.A., ed., *The Algebraic Theory of Machines, Languages, and Semigroups*. Academic Press, 1968.

[MT89] McKenzie P., Thérien D., *Automata theory meets circuit complexity.* Proceedings of the 16th ICALP, Springer Lecture Notes in Computer Science 372 (1989), 589-602.

[Ol69] Olsen J.E., *A combinatorial problem on finite Abelian groups, I.* J. of Number Theory 1 (1969), 8-10.

[Ol69a] Olsen J.E., *A combinatorial problem on finite Abelian groups, II.* J. of Number Theory 1 (1969), 195-199.

[Ra87] Razborov A.A., *Lower bounds for the size of circuits of bounded depth with basis {&, ⊕}.* Mathematicheskie Zametki 41:4 (April 1987), 598-607 (in Russian). English translation Math. Notes Acad. Sci. USSR 41:4 (Sept. 1987), 333-338.

[Sh38] Shannon C.E., *A symbolic analysis of relay and switching circuits.* Trans. AIEE 57 (1938), 713-723.

[SV81] Skyum S., Valiant L.G., *A complexity theory based on Boolean algebra.* Proceedings of the 22nd IEEE FOCS Symposium (1981), 244-253.

[Sm87] Smolensky R., *Algebraic methods in the theory of lower bounds for Boolean circuit complexity.* Proceedings of the 19th ACM STOC (1987), 77-82.

[Sm90] Smolensky R., *On interpolation by analytic functions with special properties and some weak lower bounds on the size of circuits with symmetric gates.* Proceedings of the 31st IEEE FOCS Symposium (1990), 628-631.

[St91] Straubing H., *Constant-depth periodic circuits.* International Journal of Algebra and Computation 1:1 (1991), 1-39.

[Sz90] Szegedy M., *Functions with bounded symmetric communication complexity and circuits with mod m gates.* Proceedings of the 22nd ACM STOC (1990), 278-286.

[Th91] Thérien D., *Linear lower bound on the size of $CC_2^0(q)$-circuits computing the AND function.* Manuscript (May 1991), McGill University.

[Ya90] Yao A.C.C., *On ACC and threshold circuits.* Proceedings of the 31st IEEE FOCS Symposium (1990), 619-627.

Symmetric Functions in AC^0 can be Computed in Constant Depth with Very Small Size

Ingo Wegener [*][†]   *Norbert Wurm* [*]   *Sang-Zin Yi* [*]

Abstract

It is well known that symmetric Boolean functions can be computed by constant depth, polynomial size, unbounded fan-in circuits, i.e. are contained in the complexity class AC^0, if and only if no jump in the value vector has a non polylogarithmic distance from one boundary of the value vector. This result is sharpened. Symmetric Boolean functions in AC^0 can be computed by unbounded fan-in circuits with small constant depth, an almost linear number of wires ($n \log^{O(1)} n$) and a subpolynomial (but superpolylogarithmic) number of gates ($2^{O(\log^\delta n)}$) for some $\delta < 1$).

1. Introduction

Symmetric functions form an important subclass of Boolean functions including all kinds of counting functions. A Boolean function $f : \{0,1\}^n \to \{0,1\}$ is called *symmetric* if $f(x_1, \ldots, x_n)$ depends on the input only via $x_1 + \ldots + x_n$, the number of ones in the input. Hence, a symmetric function f can be described by its *value vector* $v(f) = (v_0, \ldots, v_n)$, where v_i is the output of f on inputs with exactly i ones.

It is a classical result of Boolean complexity theory that all symmetric Boolean functions are contained in NC^1. They can even be computed by fan-in 2 circuits of logarithmic depth and linear size. We are interested in the more massively parallel, unbounded fan-in circuits. AC^0 is the class of Boolean functions computable by unbounded fan-in circuits of constant depth and polynomial size. The following theorem ([BW 87]) and [Mor 87]) is based

---

[*]FB Informatik, LS II, Univ. Dortmund, Postfach 500 500, 4600 Dortmund 50, Fed. Rep. of Germany.
[†]This author was partially supported by the DFG grant No. We 1066/2-2.

on earlier lower bounds ([Bop 86],[Yao 85],[Hås 86]) and upper bounds ([AB 84],[DGS 86],[FKPS 85]).

Theorem 1 *A sequence of symmetric Boolean functions $f = (f_n)$ with value vectors $v(f_n) = (v_0^n, \ldots, v_n^n)$ is contained in AC^0 iff $v_{g(n)}^n = \ldots = v_{n-g(n)}^n$ for some polylogarithmic function g, i.e. $g(n) = O(\log^k n)$ for some k.*

This theorem does not answer whether the symmetric functions in AC^0 can be computed by AC^0-circuits which are efficient compared with the well-known logarithmic depth, linear size, fan-in 2 circuits for symmetric functions. A similar question has been asked for adders. The carry-look-ahead method leads to an unbounded fan-in circuit with optimal depth 3 (adders of depth 2 need exponential size), $\Theta(n^2)$ gates and $\Theta(n^3)$ wires. But these adders cannot compete with the well-known fan-in 2 adders of linear size and logarithmic depth. The problem of determining the complexity of unbounded fan-in constant depth adders has been solved (see [CFS 83] for the upper bounds and [DDPW 83] for the lower bounds). For any recursive function $g(n)$ where $g(n) \to \infty$ as $n \to \infty$ there are constant depth adders of size $O(ng(n))$ but there do not exist linear size adders of constant depth. The results of this paper go in the same direction.

It is proved that symmetric Boolean functions in AC^0 can each be computed by unbounded fan-in circuits with constant depth, an almost linear number of wires $(n \log^{O(1)} n)$ and a subpolynomial but superlogarithmic number of gates $(2^{O(\log^\delta n)}$ for some $\delta < 1)$.

In Section 2 we review some known results and discuss some lower bounds. In Section 3 we reformulate the method of Denenberg, Gurevich and Shelah [DGS 86] on which our circuit design presented in Section 4 is based.

2.   Known results and simple lower bounds

Many of the known results are stated only for threshold functions which form some kind of a basis for all symmetric functions. The *threshold function* T_k^n on n variables computes 1 exactly on those inputs with at least k ones, i.e. $v_l(T_k^n) = 1$ iff $l \geq k$. $NT_k^n := \neg T_{k+1}^n$, is the corresponding *negative threshold function,* $v_l(NT_k^n) = 1$ iff $l \leq k$. $E_k^n := T_k^n \wedge NT_k^n$ is called the *exactly function,* $v_l(E_k^n) = 1$ iff $l = k$. A symmetric function f is obviously the disjunction of all E_k^n where $v_k(f) = 1$.

We have already mentioned that the upper bounds of Theorem 1 have been proved independently in three papers. Different methods have been used. For

T_k^n where $k = \lfloor \log^m n \rfloor$ Ajtai and Ben-Or ([AB 84]) proved the existence of circuits of depth $2m + 3$ and size $\Theta(n^{2m+4} \log^{m+1} n)$. The circuit whose existence has been proved by Fagin, Klawe, Pippenger and Stockmeyer ([FKPS 85]) also has depth $2m + 3$ but the size (number of gates) is larger, namely $n^{\Theta(m^2)}$. Denenberg, Gurevich and Shelah ([DGS 85]) described a uniform design of AC^0-circuits, while the other designs were nonuniform. They were only interested in the qualitative result that the circuits are AC^0-circuits. A direct implementation leads to circuits which are less efficient with respect to depth and size compared with the other circuits. Our circuit design uses the main idea of [DGS 86], the efficient coding of the cardinality of small subsets of $\{0, \ldots, n-1\}$ by short 0-1-vectors, (see Lemma 1). Since we work directly with circuits and do not use the notation of logics, we are not concerned with the size of the "universe". This simplifies our approach. Furthermore, we present an iterative circuit design and use some implementation tricks on the first levels of the circuit. This leads to the uniform design of monotone circuits for T_k^n, where $k = \lfloor \log^m n \rfloor$, with the following characteristics. The depth is $2m + 2$, the number of gates is $2^{O((\log^{m/(m+1)} n) \log \log n)}$ and, hence, $o(n^\alpha)$ for all $\alpha > 0$ but not polylogarithmic, and the number of wires equals $O(n \log^{2m+2} n)$ which is almost but not quite quasilinear $(n \log^{0(1)} n)$. Altogether we get the best known circuits with respect to all resource bounds.

Independent from our approach and with some other methods Newman, Ragde and Wigderson ([NRW 90]) have worked in the same direction. They have designed uniform circuits of depth $O(m)$, approximately $6m$, number of gates $O(n)$ and number of wires $O(n \log^{2m} n)$. Because of their use of hash functions the circuits are not monotone. This design beats our bounds only for the number of wires and is worse otherwise.

For constant k better results are possible. Friedman ([Fri 86]) has investigated the formula size of threshold functions. Using his methods the following theorem can be proved in a straightforward way.

Theorem 2 *For constant k, T_k^n can be computed by unbounded fan-in circuits of depth 3 with $O(\log n)$ gates and $O(n \log n)$ wires.*

In order to appreciate the new upper bounds we discuss some lower bounds. Each Boolean function can be computed with depth 2 by its DNF. But what is the minimal depth of circuits with polynomial size ? And what is the minimal size if the depth is bounded or even unbounded ? The following lower bound has been proved by Boppana ([Bop 86]) for monotone circuits and by Håstad ([Hås 86]) in the general form.

Theorem 3 *Polynomial size, unbounded fan-in circuits for T_k^n, where $k = \lfloor \log^m n \rfloor$, have depth $d \geq m$.*

The method of [Hås 86] does not lead to large lower bounds on the size of depth cm, unbounded fan-in circuits for T_k^n, where $k = \lfloor \log^m n \rfloor$ and $c > 1$. There are a few small size lower bounds on the number of gates and wires of unbounded fan-in, unbounded depth circuits for functions in AC^0 or NC^1. Hromkovic ([Hro 85]) proved some lower bounds by a communication complexity approach and Wegener ([Weg 91]) adapted the elimination method for proving lower bounds on the complexity of the parity function. For threshold functions we only know the following simple lower bound.

Proposition 1 *Unbounded fan-in circuits for T_k^n and $1 \leq k \leq n/2$ need at least n wires and k gates.*

Proof :   Obviously, at least one wire has to leave each variable x_i. If x_i enters an $\vee$-gate or $\bar{x}_i$ enters an $\wedge$-gate, we can eliminate this gate for $x_i = 1$. Otherwise one gate can be eliminated for $x_i = 0$. This procedure can be repeated at least k times before T_k^n is replaced by a constant function. $\square$

3.   The method of Denenberg, Gurevich and Shelah

We reformulate the method of Denenberg, Gurevich and Shelah ([DGS 86]) in a generalized form. We use a representation supporting our circuit design. The method is based on a number theoretic theorem allowing a succinct coding of the cardinality of small subsets of $\{0, \ldots, n-1\}$. Let $\mathrm{res}(i,j) := (i \bmod j) \in \{0, \ldots, j-1\}$.

Lemma 1 *Let $L := L(n)$. For sufficiently large n one can choose for each small subset S of*
$\{0, \ldots, n-1\}$, *i.e., $|S| \leq L$, some number $u < L^2 \log n$ such that $\mathrm{res}(i, u) \neq \mathrm{res}(j, u)$ for all different $i, j \in S$.*

Proof :   The proof relies on the prime number theorem in the following form.

$$\lim_{x \to \infty} \psi(x)/x = 1 \quad \text{for} \quad \psi(x) := \sum_{p\,\mathrm{prime}, p^k \leq x < p^{k+1}} \ln p^k.$$

Let $S \subseteq \{0, \ldots, n-1\}$ be some set with $|S| \leq L$. Let u be the smallest number such that $\mathrm{res}(i, u) \neq \mathrm{res}(j, u)$ for all different $i, j \in S$. From the above $\psi(u-1)$ must be greater than $(u-1)\ln 2$ for sufficiently large u. If n is large enough, then either $u < L^2 \log n$ or u will be this large.

Let a be the least common multiple of all $|i-j|$, where $i, j \in S$ and $i \neq j$, and let b be the product of all p^k where p is prime and $p^k \leq u-1 < p^{k+1}$. By definition of u for any such p^k, $\mathrm{res}(i, p^k) = \mathrm{res}(j, p^k)$ for some different $i, j \in S$. Hence, $|i-j|$ is a multiple of p^k. This implies that p^k and, since we may choose any p^k, also b divides a. Hence, $b \leq a$. Also, $a < n^{L(L-1)/2}$, since, by assumption, S has at most $\binom{L}{2}$ subsets $\{i, j\}$ where $i \neq j$ and $|i-j| < n$ for $i, j \in S$. By definition $\ln b = \psi(u-1)$.

Combining all our inequalities we have $(u-1)\ln 2 < \psi(u-1) = \ln b \leq \ln a < L^2 \ln n$ and so $u < L^2 \log n$. $\qquad\square$

This lemma can be applied several times. For the second application n is replaced by $n' := L^2 \log n$ and S is replaced by $\mathrm{res}(S, u) = \{\mathrm{res}(j, u) | j \in S\}$. It should be emphasized that $u = u(S)$ depends on S. For our circuit design this repeated application of the lemma is only of limited use. In order to keep the depth of the circuit small, we work with $L = \log^\delta n$ for some $\delta < 1$ but $\delta \approx 1$. After the first application of the lemma u can be estimated by $\log^{1+2\delta} n$ and the upper bound for u does not become smaller than $(\log^{2\delta} n) \log \log n$.

We shall see that large L corresponds to large size and small depth, and small L corresponds to small size and large depth. For the size of the circuit the function L^L is important. If $L = O(\log^\delta n)$ for some $\delta < 1$, then $L^L = 2^{O((\log^\delta n) \log \log n)} = o(n^\alpha)$ for all $\alpha > 0$. If $L = \Omega(\log n)$, then L^L grows superpolynomially.

4. The construction of small depth, small size circuits

We start with a simple but important subcircuit.

Lemma 2 NT_1^n can be computed by a circuit of depth 3 with $3\lceil \log n \rceil + 1$ gates and $(n+3)\lceil \log n \rceil$ wires. The output gate of the circuit is an $\wedge$-gate.

Proof : We assume that $n = 2^k$, otherwise one may consider NT_1^N where $N = 2^{\lceil \log n \rceil}$ and may replace $N - n$ inputs by zeros. NT_1^n is the conjunction of all prime clauses $\bar{x}_i \vee \bar{x}_j$, $i \neq j$. We use a so-called separating system to compute these prime clauses. By $(\bar{x}_0 \wedge \ldots \wedge \bar{x}_{n/2-1}) \vee (\bar{x}_{n/2} \wedge \ldots \wedge \bar{x}_{n-1})$ we compute with 3 gates and $n+2$ wires the conjunction of all prime clauses $\bar{x}_i \vee \bar{x}_j$ where the first bit of the k-bit number i equals 0 while the first bit of j equals 1. The same can be done for the other $k-1$ bits of the numbers

$0, \ldots, n-1$. Finally, the k outputs of the $\lor$-gates are combined by an $\land$-gate.

$\square$

If we would like to decrease the depth to 2, $\Theta(n^2)$ gates and wires are necessary and sufficient.

We would like to design efficient circuits for symmetric functions in AC^0. Since f and $\neg f$ have the same complexity, by Theorem 1 it is sufficient to consider symmetric functions f where $v_k(f) = 1$ only for some k where k or $n - k$ is bounded by a polylogarithmic function. These functions are the disjunction of a function f_1, where $v_k(f_1) = 1$ only for some k bounded by a polylogarithmic function, and a function f_2, where $v_k(f_2) = 1$ only for some k where $n - k$ is bounded by a polylogarithmic function. By duality, we may restrict ourselves to functions of the first type. The disjunction $f = f_1 \lor f_2$ does not increase the depth, since, by duality, exactly one of the functions f_1 and f_2 will be computed at an $\land$-gate. By the law of distributivity, we can save the last level without increasing the size of the circuit.

Let $L = L(n)$ be a function specified later and let f be a symmetric function where $v_k(f) = 1$ only for some $k \leq L$.

The first step of our circuit design is an application of the coding lemma. For all $u \in \{1, \ldots, U-1\}$ where $U := \lceil L^2 \log n \rceil$ we uniformly compute in parallel the following information:

$$y^u = (y_o^u, \ldots, y_{u-1}^u), \bar{y}^u = (\bar{y}_o^u, \ldots, \bar{y}_{u-1}^u), c^u.$$

Here $y_j^u := 1$ iff there exists an $i \in S$ for which $j = \text{res}(i, u)$, $\bar{y}_j^u$ is the negation of y_j^u, and c^u is the so-called *validity bit*, i.e. $c^u := 1$ iff for all $j \in \{0, \ldots, u-1\}$ there is at most one i such that $x_i = 1$ and $\text{res}(i, u) = j$.

Altogether we have replaced the n inputs by less than $2U^2 + U$ "new inputs". If $c^u = 1$, y^u contains the same number of ones as x. Since symmetric functions depend only on the number of ones in the input and not on their positions, y^u is in this case a valid encoding of the input. By Lemma 1 there exists at least one valid encoding if $f(x) = 1$.

The computation of $y^u, \bar{y}^u$ and c^u is easy. Let $A(j, u)$ be the set of all i where $\text{res}(i, u) = j$. Then y_j^u is the disjunction of all x_i, $i \in A(j, u)$, and $\bar{y}_j^u$ is the conjunction of all $\bar{x}_i$, $i \in A(j, u)$. y^u and $\bar{y}^u$ can be computed in depth 1 with $2n$ wires and $2u$ gates. The validity bit c^u is the conjunction of $b_o^u, \ldots, b_{u-1}^u$ where b_j^u is the NT_1-function for the variables $x_i, i \in A(j, u)$. Since the NT_1-circuits of Lemma 2 compute the output at an $\land$-gate, the conjunction for the computation of c^u can be combined with these $\land$-gates. Let $n_j := |A(j, u)|$. Then $n_o + \ldots + n_{u-1} = n$. Hence, c^u can be computed in depth 3 (last gate is an $\wedge$-gate) with

$$3 \sum_{0 \le j \le u-1} \lceil \log n_j \rceil + 1 = O(U \log n)$$

gates and

$$\sum_{0 \le j \le u-1} (n_j + 3) \lceil \log n_j \rceil = O(n \log n)$$

wires.

Hence, all $y^u, \bar{y}^u$ and c^u can be computed in depth 3 with $O(U^2 \log n)$ gates and $O(nU \log n)$ wires, and the output gate is an $\vee$-gate.

Let $v = (v_0, \ldots, v_n)$ be the value vector of f. We consider the symmetric function f^u on u variables with value vector $v^u := (v_0, \ldots, v_u)$. We know that $f(x) = f^u(y^u)$, if $c^u = 1$. Furthermore, $f(x) = 0$, if $c^u = 0$ for all u. Hence,

$$f(x) = \bigvee_{1 \le u < U} c^u \wedge f^u(y^u).$$

We compute $f^u(y^u)$ by its conjunctive normal form, i.e. in depth 2, from y^u and $\bar{y}^u$. Hence, $f^u(y^u)$ is computed on the third level by an $\wedge$-gate. Therefore, also $c^u \wedge f^u(y^u)$ can be computed on the third level. Finally, f is computed in depth 4. We remark that our circuit here ceases to be a formula.

We still have to estimate the size of the conjunctive normal forms. Since, $v_i = 0$ for $i > L$, $\binom{u}{L+1}$ clauses are sufficient to cover the inputs with more than L ones. If $v_i = 0$ and $i \le L$, $\binom{u}{i}$ clauses are sufficient to cover the inputs with i ones. Since $\binom{U}{j} \le U^j$, we can compute all $c^u \wedge f^u(y^u)$ from $y^u, \bar{y}^u$ and c^u with $O(U^{L+2}) = O(L^{2L+4} \log^{L+2} n)$ gates and $O(U^{L+3}) = O(L^{2+6} \log^{L+3} n)$ wires. We have proved the following theorem.

Theorem 4 *i) Symmetric functions f on n variables where $v_k(f) = 1$ only for some $k \le L$ can be computed by an unbounded fan-in circuit of depth 4 with $O(L^{2L+4} \log^{L+2} n)$ gates and $O(L^{2L+6} \log^{L+3} n + n\, L^2 \log^2 n)$ wires.*
ii) If $L = L(n) = O(\log^\delta n)$ for some $\delta < 1$, the number of gates is bounded by $2^{O(\log^\delta n \log\log n)}$ and the number of wires by $O(n \log^{2+2\delta} n)$.

We make some remarks. The number of gates (in part ii) of the theorem) is subpolynomial, i.e. $o(n^\alpha)$ for each $\alpha > 0$, but superpolylogarithmic. The upper bound is superpolynomial if $L = \Omega(\log n)$. We leave it to the reader to discuss functions L where $L = o(\log n)$ but $L = \omega(\log^\delta n)$ for all $\delta < 1$. We also leave it to the reader to design circuits where Lemma 1 is applied more than once.

The threshold functions are of particular interest. Our circuit works for NT^n_{k-1} only with the negative variables $\bar{x}_i$. The conjunctive normal form for $f^u = NT^u_{k-1}$ consists only of the $\binom{n}{k}$ prime clauses containing only $n-k$ negative y^u-variables each. By Lemma 1, the NT_1-circuits work only with negative variables. By construction, $\bar{y}^u_j$ is the conjunction of some negative variables. Hence, we do not need the positive y^u- and x-variables. In order to compute $T^n_k = \neg NT^n_{k-1}$ we apply de Morgan's rules to this circuit and obtain a monotone circuit for T^n_k of the same size as stated in Theorem 4.

Up to now we have not designed efficient circuits for all symmetric functions in AC^0. For the general case let us assume that $v_k = 1$ only for some $k \leq L^m$ where $L = O(\log^\delta n)$ for some $\delta < 1$ and m is a constant. For $U := \lfloor L^{2m} \log n \rfloor$ we compute as before all y^u, $\bar{y}^u$ and c^u with $O(U^2 \log n)$ gates and $O(nU \log n)$ wires. y^u and $\bar{y}^u$ are computed in depth 1 and c^u in depth 3.

Since v_k may equal 1 for $k = L^m$, the normal forms for f^u may have nonpolynomial size. We only can test y^u for up to L ones with normal forms of very small size. We do these computations for all subvectors of y^u. Afterwards we may test for L^2 ones by testing L pieces for L ones each. After m steps of this type we can test y^u for up to L^m ones. We explain these ideas now in more detail.

Let u be fixed. For all $0 \leq i \leq j \leq u-1$ we compute $s^*_1(i,j)$ which tests whether $(y^u_i, \ldots, y^u_j)$ contains exactly L ones and for all $0 \leq i \leq u-1$ and $1 \leq a \leq L$ we compute $t^*_1(i,a)$ which tests whether $(y^u_i, \ldots, y^u_{u-1})$ contains exactly a ones. These functions are computed by their conjunctive normal forms. Each of the $O(U^2)$ functions can be computed with at most $\binom{U}{L} + 1 \leq U^L + 1$ gates and at most $O(U^{L+1})$ wires. The functions s^*_1 and t^*_1 are computed on level 3. Since the last gate of the c^u-circuit has fan-in $O(U \log n)$ and is an $\wedge$-gate, we can compute on level 3 also $s_1(i,j) = c^u \wedge s^*_1(i,j)$ and $t_1(i,a) := c^u \wedge t^*_1(i,a)$ instead of s^*_1 and t^*_1. The number of wires is increased to $O(U^{L+2} \log n)$.

In the l-th step $(2 \leq l \leq m-1)$ we compute from the s_{l-1}- and t_{l-1}-functions the s_l- and t_l-functions. The function $s_l(i,j), 0 \leq i \leq j \leq u-1$, tests whether $c^u = 1$ and $(y^u_i, \ldots, y^u_j)$ contains exactly L^l ones, and $t_l(i,a), 0 \leq i \leq u-1$ and $1 \leq a \leq L^l$, tests whether $c^u = 1$ and $(y^u_i, \ldots, y^u_{u-1})$ contains exactly a ones. Obviously, $s_l(i,j) = 1$ iff $c^u = 1$ and for some $i(0) = i < i(1) < \ldots < i(L-1) < i(L) := j+1$ each of the vectors $(y^u_{i(r)}, \ldots, y^u_{i(r+1)-1})$ contains exactly L^{l-1} ones. This last information has been computed by the s_{l-1}-functions. A similar approach works for the t_l-functions. Each of the at most $U^2 + UL^l$ s_l- and t_l-functions can be computed with at most U^{L-1} gates and $U^{L-1}(L+1)$ wires.

In the m-th step we test whether $c^u = 1$ and the whole y^u-vector contains exactly k ones for some k where $v_k = 1$. Since $k \leq L^m$, $k = bL^{m-1} + a$ for some $b < L$ and $a \leq L^{m-1}$. The vector y^u contains exactly k ones iff it can be broken into $b + 1$ pieces such that each of the first b pieces contains exactly L^{m-1} ones and the last one contains exactly a ones. On this last level we combine the $\lor$-gates for the different values of k and u. We have computed $f(x)$, since all s_1 and t_1 are equal to 0, if $c^u = 0$ for all u. Furthermore, $c^u = 1$ for all u, if $x_o = \ldots = x_{n-1} = 0$.

If $m \geq 2$, the depth of the circuit equals $2m$. For the s_1- and t_1-functions depth 3 is sufficient and the last level is an $\land$-level. Each further step is done in depth 2 and the first level is an $\land$-level. Hence, levels 3 and 4 can be combined.

The number of gates can be estimated by

$$O(U^2 \log n + U^{L+2} + mU^{L+1} + UL^m U^{L-1}) = 2^{O(\log^\delta n \log \log n)}.$$

The number of bits in all $(y^u, \bar{y}^u, c^u)$ is very small. Hence, the number of wires which enter the gates on the first two levels dominates the number of all other wires. Therefore, the number of wires can be estimated by $O(nU \log n) = O(n \log^{2m\delta+2} n)$.

We have proved the following theorem.

Theorem 5 *Let $L = O(\log^\delta n)$ for some $\delta < 1$. Symmetric functions f on n variables where $v_k(f) = 1$ only for some $k \leq L^m$ and m is a constant, can be computed by unbounded fan-in circuits of depth $2m$ with $2^{O(\log^\delta n \log \log n)}$ gates and $O(n \log^{2m\delta+2} n)$ wires.*

For threshold functions T_k^n, where $k \leq L^m$, we even obtain monotone circuits. We consider again NT_{k-1}^n. We compute only $\bar{y}^u$ and c^u using only the negative variables $\bar{x}_i$. In our circuit design we have tested whether subvectors of y^u contain exactly a certain number of ones. Each "exactly" can be replaced by "at most", if we want to compute NT_{k-1}^n. Then s_1^* and t_1^* are NT-functions whose conjunctive normal forms contain only the negative variables $\bar{y}_j^u$.

In order to compare our results with the results of the other papers we consider the special case of T_k^n and $k = \lfloor \log^m n \rfloor$. Let $L := \log^{m/(m+1)} n$. Then $L^{m+1} = \log^m n$. We have proved the following theorem.

Theorem 6 *The threshold functions T_k^n where $k = \lfloor \log^m n \rfloor$ for constant m can be computed by unbounded fan-in monotone circuits of depth $2m+2$ with $2^{O((\log^{m/(m+1)} n) \log \log n)}$ gates and $O(n \log^{2m+2} n)$ wires.*

Remark Håstad, Wegener, Wurm and Yi ([HWWY 91]) have recently improved these results reducing the depth to $\lfloor m \rfloor + 1$ without increasing the size.

References

[AB 84]   Ajtai M., Ben-Or M., A theorem on probabilistic constant depth computations. 16th ACM Symp. on Theory of Computing, 471-474, 1984.

[Bop 86]   Boppana R., Threshold functions and bounded depth monotone circuits. Journal of Computer and System Sciences 32, 222-229, 1986.

[BW 87]   Brustmann B., Wegener I., The complexity of symmetric functions in bounded-depth circuits. Information Processing Letters 25, 217-219, 1987.

[CFL 83]   Chandra A., Fortune S., Lipton R.J., Unbounded fan-in circuits and associative functions. 15th ACM Symp. on Theory of Computing, 52-60, 1983.

[DGS 86]   Denenberg L., Gurevich Y., Shelah S., Definability by constant-depth polynomial-size circuits. Information and Control 70, 216-240, 1986.

[DDPW 83] Dolev D., Dwork C., Pippenger N.J., Wigderson A., Superconcentrators, generalizers and generalized connectors with limited depth. 15th ACM Symp. on Theory of Computing, 42-51, 1983.

[FKPS 85]   Fagin R., Klawe M.M., Pippenger N.J., Stockmeyer L., Bounded-depth, polynomial-size circuits for symmetric functions. Theoretical Computer Science 36, 239-250, 1985.

[Fri 86]   Friedman J., Constructing $O(n \log n)$ size monotone formulae for the k-th elementary symmetric polynomial of n Boolean variables. SIAM Journal on Computing 15, 641-654, 1986.

[Hås 86]   Håstad J., Almost optimal lower bounds for small depth circuits. 18th ACM Symp. on Theory of Computing, 6-20, 1986.

[HWWY 91] Håstad J., Wegener I., Wurm N., Yi S.-Z., Optimal depth, very small size circuits for symmetric functions in AC^0. To appear: Information and Computation.

[Hro 85]   Hromkovic J., Linear lower bounds on unbounded fan-in Boolean circuits. Information Processing Letters 21, 71-74, 1985.

[Mor 87]   Moran S., Generalized lower bounds derived from Håstad's main lemma. Information Processing Letters 25, 383-388, 1987.

[NRW 90]   Newman I., Ragde P., Wigderson A., Perfect hashing, graph entropy and circuit complexity. Proc. 5th Structure in Complexity Theory, 91-99, 1990.

[Weg 91]   Wegener I., The complexity of the parity function in unbounded fan-in, unbounded depth circuits. Theoretical Computer Science 85, 155-170, 1991.

[Yao 85]   Yao A.C., Separating the polynomial-time hierarchy by oracles. 26th IEEE Symp. on Foundations of Computer Science, 1-10, 1985.

Boolean Complexity and Probabilistic Constructions

Miklós Ajtai *

Abstract

The main result is a theorem which states, that in probabilistic constructions the independence of random variables in certain cases can be replaced by a form of weak dependence, where the weakness is characterized by the circuit complexity of the dependence. The result makes random constructions applicable for a larger class of problems.

Assume that we pick t elements at random and with uniform distribution from the finite set A. Let K be the set formed by these elements. If we pick also a subset S of A with an arbitrary distribution but independently from the K, then it is trivially true that with a high probability either the density of S in A is small, or S contains at least one point from K. E.g., with a probability of at least $1 - e^{t^{1-\epsilon}}$ we have that either $|S| < t^{-\epsilon}|A|$ or S and K have a common point.

This statement does not remain true if we drop the condition that S is independent from K, e.g., S can be the complement of K in the set A. However, if for all $a \in A$, there is a not too large, constant depth Boolean circuit, which given the set $K - a$ as an input decides whether a is in S, then with a probability of at least $1 - 2^{-n^\epsilon}$, either $|S| < n^{-\epsilon}|A|$ or S and K has a common element.

1. Introduction

The main aim of this paper is to prove a theorem that we get from a trivial but frequently used fact about independent random variables, replacing the notion of independence by a weak form of Boolean dependence. (For a simple formulation of a special case of the theorem see Theorem 1.1 below.)

The mentioned trivial fact about independent random variables is the following:

---

*IBM Almaden Research Center, K/53, 650 Harry Road, San Jose, CA., 95120, USA

Suppose that we take t random points independently and with uniform distribution from the set A. If a subset S of A is given so that the density of S is not too small, say greater than $t^{-\varepsilon}$, then with a high probability at least one of our selected points will be in S. This remains true even if we choose the set S at random too, with arbitrary distribution, but independently of the t points. If we denote by K the set formed by the randomly selected points then we have that

$$(1.1) \qquad P(|S| \leq t^{-\varepsilon}|A| \text{ or } K \cap S \neq \emptyset)$$

is very close to 1, (at least $1 - e^{-t^{1-\varepsilon}}$), that is with a high probability either $|S|$ is small or S contains a point from K.

If we do not require the set S to be independent of K then the analogous statement is false, in fact if we define S as the complement of K in the set A then the probability in (1.1) is 0 (provided that A has more than $2t$ elements). The essential feature of this counterexample is the following. Suppose that we define S from K, by deciding about every point $x \in A$ whether it will be in S. When we decide about the point x we know already whether x is in K or not. In the above example we simply said that $x \in S$ iff $x \notin K$. We may avoid this type of counterexample by requiring that when we decide about x we do not know whether $x \in K$ or $x \notin K$, we know only the set $K - \{x\}$.

More precisely assume that for each $x \in A$ a unary relation R_x is given on the set of all subsets of $A - \{x\}$. If K is a subset of A let $S = \{x \in A| R_x(K - \{x\})\}$. Since $x \in S$ does not depend on the information whether $x \in K$, we may hope that if K is the set formed by our randomly selected t points, then with high probability either $|S|$ is small or S contains a point from K. Unfortunately this is not true. Actually the probability in (1.1) can be $1/2$. Indeed for $i = 0, 1$ let $R_x^{(i)}$ be the relation defined by $R_x^{(i)}(H)$ iff $|H| \equiv i \pmod 2$. Let $S^{(i)}$, $i = 0, 1$ be the set S defined from K using the relations $R_x^{(i)}$. Suppose now that we pick a random K. There is an $i \in \{0, 1\}$ so that with a probability of at least $1/2$ the parity of $|K|$ is the same as the parity of i. If such an i and K is fixed then $S^{(i)}$ is the complement of K, and so $S^{(i)} \cap K = \emptyset$ and $|S^{(i)}| = |A - K| \geq t^{-\varepsilon}|A|$ if $|A| > 2t$.

This counterexample remains also valid for K being a random subset of A with exactly t elements, or K being randomized by deciding about each element of A independently with a probability of $t/|A|$ whether it will be in K. The choice of the random variable K from the three mentioned possibilities will have no importance in the following discussion either. Still for the sake of definiteness we assume that K is the originally defined random variable, that is, the set formed by t random points picked independently and with uniform distribution from A. Since the random points may coincide it is possible that $|K| < t$.

The crucial part of the previous counterexample was that if we know something about the cardinality of $K - \{x\}$ (the parity) then it may give some nontrivial information about the question $x \in K$? However if we suppose that the relations R_x can be computed by a not-too-large Boolean circuit of constant depth, then knowing that it is difficult to compute the cardinality of a set or even to get some nontrivial information about it (like parity) with such circuits, we may expect to avoid counterexamples of the previous type.

We want to formulate a theorem of the following type: Suppose that K is the random variable described above and for each possible value of K we define a set $S \subseteq A$ in the following way. Assume that for each $x \in A$ a Boolean circuit C_x is given with $|A| - 1$ inputs, 1 output, the depth of the circuit is constant and there are some restrictions on the size and structure of the circuit. We will associate each input of the circuit C_x with an element of $A - \{x\}$. The set $K - \{x\} \subseteq A - \{x\}$ can be considered as an input vector $\vec{y}_{K-\{x\}}$ of the circuit C_x namely the value at an input associated with the point $z \in A - \{x\}$ will be 1 iff $z \in K - \{x\}$.

We will prove a theorem saying that if all of the circuits $C_x, x \in A$ meet certain requirements and $S = \{x \in A|\ C_x(\vec{y}_{K-\{x\}}) = 1\}$ then

(1.2)   with a probability of at least $1 - 2^{-t^{\varepsilon}}$ either $|S| \leq t^{-\varepsilon}|A|$ or $S \cap K \neq \emptyset$.

First we formulate the theorem in the special case when $|A| \leq t^c$.

Theorem 1.1 For all $c > 0, d > 0$, there exists $\varepsilon > 0$ so that if t is sufficiently large, A is a set, $|A| \leq t^c$, and for each $a \in A$, C_a is a Boolean circuit of depth at most d and of size at most $2^{t^{\varepsilon}}$ with $|A| - 1$ inputs (each input sequence is associated with a subset of $A - \{a\}$), then the following holds: if K is the set of t random points taken independently and with uniform distribution from A, and $S = \{x \in A \mid C_x(K - \{x\}) = 1\}$, then with a probability of at least $1 - 2^{-t^{\varepsilon}}$ we have that either $|S| \leq t^{-\varepsilon}|A|$ or $S \cap K$ is nonempty.

Before we describe the properties required from the circuits C_x in the general case (with no restriction on the size of A) we give a simple example. Suppose that a directed graph G with no loops is given on the vertex set A. Let S be the set of those points x which are connected to every point b of K by a directed edge from b to x. Then there is an absolute constant $\varepsilon > 0$ so that if t is sufficiently large then 1.2 holds, that is, with high probability either only a small number of points can be reached simultaneously from every points of

K by a directed edge or there is a point in K which can be reached from every other points in K by a directed edge. (This statement can be proved directly without using the general theorem formulated below.) We show what the circuits C_x are in this case. If for each $a \in A$, y_a is a Boolean variable whose value is 1 iff $a \in K$ then we may define the circuit C_x by $C_x = \bigwedge_{x \neq b, \langle b,x \rangle \notin G} \neg y_b$.

It is easier to describe the more general result in terms of formulae instead of circuits. In our example about the directed graph, the set S can be defined in the following way: $x \in S$ iff $\forall y \in K - \{x\}$ $\langle y,x \rangle \in G$. That is, here we have a first order formula on the structure $\langle A, G \rangle$ where the quantifier is restricted to the set $K - \{x\}$. As we will see, (1.2) remains true if S is defined by a formula of this type (whose length is a constant, i.e. does not depend on t), even if we allow to quantification over subsets of K of size at most t^ε. We may also allow quantification from a fixed set Γ with at most 2^{t^ε} elements.

In the formulation of our theorem we will use the following notation: if H is a set and $u \geq 0$ is a real number then $\lceil H \rceil^u$ will denote the set of all subsets of H with at most u elements. If H, G are sets then $H \times G$ is the set of all ordered pairs whose first element is in H and whose second is in G.

Theorem 1.2 *For each positive integer d there is an $\varepsilon > 0$ so that if t is sufficiently large, Γ, A are finite sets, $|\Gamma| \leq 2^{t^\varepsilon}$ and for all $x \in A$, Ψ_x is a relation on the set of sequences whose first d elements are in Γ and second d elements are subsets of A with at most t^ε elements, then the following holds: K is the set of t random points taken from A independently and with uniform distribution and S is the set of all $a \in A$ with the property*

$$(1.3) \quad \exists X_1 \in \lceil K - \{a\} \rceil^{t^\varepsilon}, \gamma_1 \in \Gamma, \forall X_2 \in \lceil K - \{a\} \rceil^{t^\varepsilon}, \gamma_2 \in \Gamma, \ldots$$
$$\ldots \mathcal{Q} X_d \in \lceil K - \{a\} \rceil^{t^\varepsilon}, \gamma_d \in \Gamma, \Psi_a(\gamma_1, \ldots, \gamma_d, X_1, \ldots, X_d),$$

(where $\mathcal{Q}$ is the quantifier $\forall$ if d is even and the quantifier $\exists$ otherwise), then with a probability of at least $1 - 2^{-t^\varepsilon}$ we have that either $|S| \leq t^{-\varepsilon}|A|$ or $S \cap K$ is nonempty.

In the proof we show that if we write K in the form of $J \cup L$ where J is the set formed by the first $t - \lceil t^{\varepsilon'} \rceil$ points and L is the set formed by the remaining $\lceil t^{\varepsilon'} \rceil$ points, and first we randomize only J then we get a set so that with a high probability the defining formula of S (now only as a function of L) will be equivalent to a Σ_2 formula, that is a formula of the same type with $d = 2$, or to a Π_2 formula (what we get from the Σ_2 formula by reversing the order of quantification). This is Theorem 1.3.

We will use the following notation in the statement of Theorem 1.3. If D, A are sets then $\prod_D A$ will denote the set of all functions whose domain is D and whose range is a subset of A. If f is a function defined on the set D then range$(f) = \{y | \exists x \in D \; y = f(x)\}$. If A, Γ are sets and u is a positive real number then $\langle\!\langle A \rangle\!\rangle_\Gamma^u$ will denote the set of all pairs $\langle \gamma, X \rangle$ where $\gamma \in \Gamma$ and $X \in \lceil A \rceil^u$.

Theorem 1.3 *For all positive integers d there exists an $\varepsilon > 0$ so that for all $\mu > 0$ there exist $\lambda > 0, c > 0$ so that if t is sufficiently large then the following holds:*

Suppose that Γ, A, H, G are finite sets $|\Gamma| \le 2^{t^c}$, $|G \cup H| = t$, $G \cap H = \emptyset$, $|G| \le t^\varepsilon$, Ψ is a d-ary relation on $\langle\!\langle A \rangle\!\rangle_\Gamma^{t^\lambda}$. Then for each $Z \subseteq A$ there exists a binary relation R_Z on $\lceil A \rceil^{t^\mu}$ so that if f is a random element of $\prod_{H \cup G} A$ with uniform distribution, range$(f) = K$, range$(f|_H) = J$ and range$(f|_G) = L$, then with a probability of at least $1 - 2^{-t^c}$ the following holds:

(1.4)

$$\forall X_1 \in \langle\!\langle K \rangle\!\rangle_\Gamma^{t^\lambda} \exists X_2 \in \langle\!\langle K \rangle\!\rangle_\Gamma^{t^\lambda} ... \mathcal{Q} X_d \in \langle\!\langle K \rangle\!\rangle_\Gamma^{t^\lambda} \; \Psi(X_1, ..., X_d) \equiv$$

$$\exists U \in \lceil L \rceil^{t^\mu} \forall V \in \lceil L \rceil^{t^\mu} R_J(U, V) \equiv \forall V \in \lceil L \rceil^{t^\mu} \exists U \in \lceil L \rceil^{t^\mu} R_J(U, V),$$

where $\mathcal{Q}$ is the quantifier $\exists$ if d is even otherwise it is the quantifier $\forall$.

This theorem shows some similarity to the "circuit collapsing" results used to prove lower bounds for parity circuits. We do not use in our proofs any results of this type but the reader who is familiar with some of these lower bound proofs may think of them as a motivation for the present approach. Using the methods given in [Aj] or by Furst, Saxe and Sipser in [FSS] we can prove that if we randomly assign values to the input variables of a constant depth polynomial size circuit, then with high probability the circuit collapses into a circuit which depends only on a constant number of variables. An explicit statement and proof of this fact is given by Ajtai and Wigderson in [AW]. In the language of formulae we can say that the formula collapses into a Σ_0 formula. The proofs of Yao ([Y]), and Håstad ([H]) are applicable to a larger class of circuits, but here the circuits are collapsing into circuits of more complicated types (disjunctions of conjunctions). Formally this is the most similar case to our present setup, but we were not able to use their results in our proof.

We describe a consequence of Theorem 1.3 for the $|A| \le t^c$ special case. This can be considered as an analogue of the mentioned results of Yao and Håstad for a different type of randomization. (This statement also can be used to give a 2^{n^ε} lower bound on the size of constant depth parity circuits with n

inputs. Actually Theorem 1.1 also yields a similar lower bound using the counterexample based on the parity of $|K - \{x\}|$.)

The mentioned collapsing results are of the following form: we have a Boolean circuit with n inputs with certain restrictions on the depth and the size. We give random values 0 and 1 to a random set of the inputs and leave the other inputs undecided. The theorems state that with high probability in the remaining undecided variables the circuit is simpler (e.g. depth 2). Now we will use a different type of randomization. We imagine that originally all of the t^c input variables have value 0 and we randomize t of them (not necessarily distinct) that will be changed into 1. It is a consequence of Theorem 1.3 that after the randomization of $t - t^e$ of the t variables for the remaining randomization the circuit is equivalent to a conjunction of disjunctions of conjunctions, that is, a depth 3 circuit of the type $\wedge \vee \wedge$:

Corollary 1.4 . *For all positive integers d, c there exists an $\varepsilon > 0$ so that for all $\mu > 0$ there exist $\lambda > 0, c' > 0$ so that if t is sufficiently large then the following holds:*

Suppose that C is a Boolean circuit of depth at most d and size at most 2^{t^λ} with $[t^c]$ input variables x_a, $a \in A$, and $|G \cup H| = t$, $G \cap H = \emptyset$, $|G| \leq t^\varepsilon$. Then for each $Z \subseteq A$ there exists a Boolean circuit C_Z of size at most 2^{t^μ} with the input variables x_a, $a \in A - Z$, which is a conjunction of disjunctions of conjunctions so that if f is a random element of $\prod_{H \cup G} A$ with uniform distribution, $\mathrm{range}(f) = K$, $\mathrm{range}(f|_H) = J$, $\mathrm{range}(f|_G) = L$ and $g = f|_G$, then with a probability of at least $1 - 2^{-t^{c'}}$ the following holds:

$$C(v_f^A) \equiv C_J(v_g^{A-J})$$

where v_h^Y is the evaluation of the variables x_a, $a \in Y$ defined by "$x_a = 1$ iff $a \in \mathrm{range}(h)$"

The proofs of both Theorem 1.5 and Theorem 1.3 are based on a combinatorial lemma. This lemma states the following: assume that for each subset B of A with at most t^λ elements, where $\lambda > 0$ is a small constant and for each $\gamma \in \Gamma$, a set $\Phi(B, \gamma)$ is given which contains subsets of A with at most t^λ elements. Suppose that K, J, L are the same as in Theorem 1.3. The lemma states that with high probability, if for all $B \subseteq \lceil K \rceil^{t^\lambda}, \gamma \in \Gamma$ there is an $X \subseteq K$ with $X \in \Phi(B, \gamma)$, then there is a single small subset M of L so that all of the sets X can be taken from $J \cup M$.

Lemma 1.5 *For all sufficiently small $\varepsilon > 0$ if $\mu > 0$ is sufficiently small with respect to ε and $\lambda > 0$ is sufficiently small with respect to μ, t is sufficiently* large, A, H, G, Γ are finite sets, $H \cap G = \emptyset$, $|G| = [t^\varepsilon]$, $|H| + |G| = t$, $|\Gamma| \le 2^{t^\lambda}$ and for each $B \in [A]^{t^\lambda}$ and $\gamma \in \Gamma$, $\Phi(B, \gamma)$ is a set containing subsets of A with no more than t^λ elements, then the following holds:

Suppose that f is taken at random with uniform distribution from $\prod_{H \cup G} A$. Then with a probability of at least $1 - 2^{-t^{\mu/2}}$ we have that:

If for each $B \subseteq \text{range}(f)$, $|B| \le t^\lambda$ and $\gamma \in \Gamma$ there is a $\varphi \subseteq \text{range}(f)$ so that $\varphi \in \Phi(B, \gamma)$, then

(**1.5**)   there is a subset M of G with at most t^μ elements so that for each $B \subseteq \text{range}(f), |B| \le t^\lambda, \gamma \in \Gamma$ there is a $\varphi \subseteq \text{range}(f|_{H \cup M}) \cup B$ with $\varphi \in \Phi(B, \gamma)$.

Application. We describe here an application of Theorem 1.2. We formulate only a theorem, the proof will be given in a separate paper.

The lower bound results for parity circuits use the following property of the set E of sequences of length n with an even number of 1's: if $x \in E$ and we change x at an arbitrary place then we get a sequence outside E. It means that in some geometric sense each of the points of E is a boundary point.

The set of $0, 1$ sequences of length n can be considered as the direct product of n sets of size 2. We will be interested in the direct product of sets of arbitrary size where each element of each set is coded by a $0, 1$ sequence, so the elements of the product space can be considered as inputs for a suitable Boolean circuit. We will show that if a subset D of the product space is definable by a not too large circuit of constant depth, then it cannot contain too many "boundary points", where b is boundary point of D if changing a random component of b at random we get a point outside D with a not too small probability. More precisely:

Definition. Assume that $D \subseteq \prod_{p \in P} X_p$, where P is a finite set and for each $p \in P$, X_p is a finite set. Suppose that n is a positive integer, $\delta > 0$ and $b \in D$. We say that b is a δ-boundary point of D if picking a random $p \in P$ with uniform distribution then changing the P component of b at random (with uniform distribution on X_p), we get a $b' \notin D$ with a probability of at least δ.

In the theorem below we will use the following notation. d, n are positive integers, $\varepsilon > 0$, $r \le 2^{n^\varepsilon}$, P is a set with at most n elements, for each $p \in P$, X_p is a set and u_p is a function defined on X_p each of whose values is a $0, 1$ sequence of length r.

If C is a circuit with $r|P|$ inputs then C defines a function on the Cartesian product $\prod_{p \in P} X_p$ by $C(\langle \ldots, x_p, \ldots \rangle) = C(\ldots, u(x_p), \ldots)$.

Theorem 1.6 *For all positive integer d, there exists an $\varepsilon > 0$ so that if n is sufficiently large, P, X_p, u_p are given with the properties described above, C is a depth d circuit of size at most 2^{n^ε} with rn inputs and $D = \{x \in \prod_{p \in P} X_p \mid C(x) = 1\}$ then the number of $n^{-\varepsilon}$-boundary points of D is at most*

$$2^{-n^\varepsilon} \prod_{p \in P} |X_p|.$$

2. Sketch of the proof.

We sketch the proof of the two mentioned theorems using Lemma 1.5. Lemma 1.5 essentially is the same as Theorem 1.3 in the $d = 2$ case. That is we have a formula of the type e.g.

$$\Theta \equiv \forall X \in \langle\!\langle K \rangle\!\rangle_\Gamma^{t^\lambda} \exists Y \in \langle\!\langle K \rangle\!\rangle_\Gamma^{t^\lambda} \Psi(X, Y)$$

and we have to show that with high probability the quantifiers can be switched, at least if we fix J, the larger part of K at random and the remaining part L takes the role of K, that is we may quantify subsets of L. For the moment let us assume that in the formula Θ, X and Y are restricted to $\lceil K \rceil^{t^\lambda}$ instead of $\langle\!\langle K \rangle\!\rangle_\Gamma^{t^\lambda}$. According to Lemma 1.5 if for all X there is a Y with $\Psi(X, Y)$ then with high probability there is a subset U of L with at most t^μ elements so that for each X there is a Y contained in $J \cup U \cup X$ with $\Psi(X, Y)$.

So our equivalent Σ_2 formula will be the following: There is a $U \subseteq L, |U| \le t^\mu$ so that for all X "there is a $Y \subseteq J \cup U \cup X$ with $\Psi(X, Y)$". The truth value of the expression in quotation marks depends only on X and U (for a fixed J), so if we define this as the relation R_J then we have $\Theta \equiv \exists U \forall X R_J(U, V)$ with high probability. The general case can be handled in a similar way.

If we accept Theorem 1.3 then the proof of Theorem 1.2 is the following. First we show that it is enough to prove Theorem 1.2 for Σ_2 formulae. Indeed Theorem 1.3 guarantees that in some sense every formula is equivalent to a Σ_2 formula. We will show by a simple probabilistic argument that if we replace all of the formulae in Theorem 1.2 by the corresponding Σ_2 formulae then we get an equivalent statement.

The proof of the theorem for Σ_2 formulae goes in the following way. Assume that for each $x \in A$, Θ_x is the corresponding Σ_2 formula. We have to prove the following. If we pick K at random then with high probability either the density of S is small in A or $\exists x \in K \; \Theta_x$. This last formula is also a Σ_2

formula. For the sake of simplicity let us assume that only subsets of K are quantified (and not the elements of Γ), that is, the formula is of the form: $\exists X \forall Y \Psi(X, Y)$. According to Lemma 1.5 if this formula is false then there is a $U \subseteq L$, $|U| \leq t^\mu$ so that for all X there is a Y with $\neg \Psi(X, Y)$ and $Y \subseteq J \cup U \cup X$. We will show that for any fixed U the probability p that the density of S is small and $\forall X \exists Y \subseteq J \cup U \cup X \ \neg \Psi(X, Y)$ is subexponentially small. It will be enough since if $U = \mathrm{range}(f|_M)$ then we have a relatively small number of choices for M. So if p_M is the probability that $\forall X \exists Y \subseteq J \cup U \cup X \neg \Psi(X, Y)$ then $p \leq \sum p_M$ will be sufficiently small.

Assume now that U (and M) are fixed. The statement of Theorem 1.2 in this case will be equivalent to the following assertion.

Lemma 2.1 *There is a $\xi > 0$ so that if t is sufficiently large, A, F are finite sets, $|F| = t$, and for all $x \in A$, Φ_x is a set of subsets of $A - \{x\}$ each containing no more than t^ξ elements, then the following holds:*

if f is picked at random with uniform distribution from $\prod_F A$ and

$\bar{S}_f = \{a \in A | \exists U \in \Phi_a \ \ U \subseteq \mathrm{range}(f)\}$ *then with a probability of at least $1 - 2^{-t^\xi}$ we have that either $|\bar{S}_f| \leq t^{-\xi}|A|$ or $|\bar{S}_f \cap \mathrm{range}(f)| > 0$.*

3. Proof of Lemma 1.5.

In the proof we will use a lemma which is, roughly speaking, a generalization of the following statement. Assume that there is a set Φ of small subsets of A and we randomly pick first t points of A then s points. If $s \leq t^\varepsilon$ and the first t points do not cover any element of Φ then it is unlikely that the $t + s$ points together will cover one.

Lemma 3.1 *There exist positive real numbers c, β so that if $\mu > \lambda > 0$ are sufficiently small and t is sufficiently large, $s \leq t^\mu$, A, H, G are finite sets, $|H| = t$, $|G| = s$, $0 \leq k < t^\lambda$, $1 \leq l < t^\lambda$, Φ is a subset of the powerset of A, then the following holds:*

If f is a random element with uniform distribution from $\prod_H A$, then with a probability of at least $1 - 2^{-t^c}$, f meets the following requirement:

(3.1) If d is a random element with uniform distribution from $\prod_G A$ then the probability of the following event is smaller than $t^{-\beta l}$:

(3.2) $\exists \varphi', \varphi'' \ |\varphi'| \leq k, |\varphi''| \geq l$, $\varphi' \subseteq \mathrm{range}(f)$, $\varphi'' \subseteq \mathrm{range}(d)$, $\varphi' \cup \varphi'' \in \Phi$ but for all $\psi'' \subseteq \mathrm{range}(d)$, $|\psi''| < l$ there is no $\psi' \subseteq \mathrm{range}(f)$ with $\psi' \cup \psi'' \in \Phi$.

Lemma 3.1 states that an event B in the direct product of two probability spaces is of small probability in some strong sense. In the lemma the two spaces are the randomization of f on H and the randomization of d on G and B is the event described in (3.2). If the first probability measure is κ defined on the probability space K, and the second is λ defined on the probability space L, then the statement of the Lemma is the following: $\kappa(\{f \in K|\ \lambda((\{f\} \times L) \cap B) < t^{-\beta l}\}) \geq 1 - 2^{-t^c}$.

This way we measure the "smallness" of the set B by two parameters $t^{-\beta l}$ and $1 - 2^{-t^c}$. Generally if γ and δ are arbitrary positive numbers and D is an arbitrary event in $K \times L$ then we will say that D is γ, δ small if

$$\kappa(\{f \in K|\ \lambda((\{f\} \times L) \cap D) < \gamma\}) \geq 1 - \delta.$$

We will need the following trivial property of this notion:

(3.3)   if $D \subseteq \cup_{i=1}^{h} F_i$ and F_i is γ_i, δ_i small for all $i = 1, ..., h$, then D is $\sum_{i=1}^{h} \gamma_i$, $\sum_{i=1}^{h} \delta_i$ small.

Using 3.3 we may easily reduce the Lemma to its $k = 0$ special case:

Proposition 3.2 *If Lemma 3.1 holds in the $k = 0$ special case, then it holds in general.*

Proof. Assume that the $k = 0$ case holds with $c \to c'$, $\beta \to \beta'$, $1/4 > \lambda > 0$, $\mu > 0$. Let $c = c'/4$, $\beta = \beta'/4$ and assume that A, H, G are given as in Lemma 3.1 and B is the event defined above. We partition H into $k + 1$ subsets $H_1, ..., H_{k+1}$ of roughly equal size, say $\frac{t}{2(k+1)} \leq |H_i| \leq \frac{2t}{k+1}$, for $i = 0, 1, ..., k$. For each $i = 0, 1, ..., k$ let F_i be the following event:

(3.4) $\exists \varphi', \varphi''\ |\varphi'| \leq k, |\varphi''| \geq l,\ \varphi' \subseteq \text{range}(f|_{H-H_i}),\ \varphi'' \subseteq \text{range}(d),$
$\varphi' \cup \varphi'' \in \Phi$ but for all $\psi'' \subseteq \text{range}(d),\ |\psi''| < l$ there is no $\psi' \subseteq \text{range}(f)$ with $\psi' \cup \psi'' \in \Phi$,

that is we get 3.4 from 3.2 by adding the requirement that φ' is already a subset of the range of f on $H - H_i$. 3.2 implies that 3.4 holds for at least one $i \in \{0, 1, ..., k\}$ that is $B \subseteq \cup_{i=0}^{k} F_i$. Indeed $|\varphi'| \leq k$ implies that the inclusion $\varphi' \subseteq \cup_{i=0}^{k}\text{range}(f|_{H_i})$ remains valid if we leave out a suitably chosen term. Therefore if we show that for all $i \in \{0, ..., k\}$, F_i is $t^{-\beta' l/2}$, $2^{-t^{c'}/2}$ small, then by (3.3) we have that B is $(k+1)t^{-\beta' l/2}$, $(k+1)2^{-t^{c'}/2}$ small. Since $k \leq t^\lambda$, $c = c'/4$, $\beta = \beta'/4$ this implies the assertion of the lemma.

Suppose that first we randomize f everywhere outside H_i and let $\bar{f}$ be the function that we get this way. Let $\Phi' = \{\varphi - \text{range}(\bar{f}) | \varphi \in \Phi\}$. Now we randomize f on H_i and d on G. For this randomization we apply the $k = 0$ case of the lemma with $H \to H_i$, $t \to |H_i|$, $\Phi \to \Phi'$. We get that with a probability of at least $1 - 2^{(t^{1-\lambda}/2)^{c'}}$ we get an f so that the probability of 3.2 is smaller than $(t^{1-\lambda}/2)^{-\beta'l}$. Since this is true for every fixed $\bar{f}$ we have that F_i is $t^{-\beta'l/2}$, $2^{-t^{c'}/2}$ small which completes the proof of the proposition.

The following lemma is a reformulation of Lemma 3.1 for $k = 0$. ($k = 0$ implies that we may consider only those sets in Φ which have exactly l elements.)

Lemma 3.3 *There exist $c > 0, \beta > 0$ such that if $\mu > \lambda > 0$ are sufficiently small, t is sufficiently large, $1 \le s \le t^\mu$, $l \le t^\lambda$ A, H, G are finite sets, $|H| = t$, $|G| = s$, Φ is a family of subsets of A, $\varphi \in \Phi \to |\varphi| = l$, and f is a random element of $\prod_H A$ with uniform distribution, then with a probability of at least $1 - 2^{-t^c}$ the following holds:*

(3.5) *if we pick a random $g \in \prod_G A$ with uniform distribution then the probability of the following event is at most $t^{-l\beta}$:*

(*) $\exists \varphi \in \Phi$ $\varphi \subseteq \text{range}(g)$ *but* $\forall D \subseteq G$, *if* $|D| < l$ *then there is no* $\psi \in \Phi$ *so that* $\psi \subseteq \text{range}(f) \cup \text{range}(g|_D)$.

Proof. It is sufficient to prove the Lemma in the case $s = l$. Indeed let U be a fixed subset of G with l elements. The statement of the Lemma for $l = s$ implies that for each fixed U there is an exceptional subset E_U of $\prod_H A$ with at most $2^{-t^c} |A|^{|H|}$ elements so that for each f outside E_U, (3.5) holds with $G \to U$. Since there are at most $(t^\mu)^{t^\lambda} = t^{\mu t^\lambda} \le 2^{t^{2\lambda}}$ choices for the set U, the union of all of the exceptional sets E_U contains at most $2^{-t^c} |A|^{|H|}$ elements. Suppose that f is not in this union. If we pick g at random then for each fixed U the probability that (*) holds with $\varphi \subseteq \text{range}(g|_U)$ is smaller than $t^{-\beta l}$ so the probability that it holds with some U is smaller than $t^{\mu l} t^{-\beta l} \le t^{-\beta' l}$.

Next we prove that we may suppose that each $\varphi \in \Phi$ meets the following requirement:

(3.6) *if H' is a subset of H with at most $t^{1/3}$ elements and f' is taken at random with uniform distribution from $\prod_{H'} A$ then the probability of the following event is at most $t^{-1/3}$:*

(**) *there is $\psi \in \Phi$ so that $\psi \subseteq \text{range}(f') \cup \varphi$ and $\text{range}(f') \cap \psi \neq \emptyset$.*

Suppose that $\varphi \in \Phi$ does not satisfy (3.6). Then, if we take f from $\prod_H A$ at random then with a probability of at least $1 - 2^{-t^{1/3}}$ we will have (**) with $f' \to f$. Indeed we may partition H into $t^{2/3}$ subsets of size $t^{1/3}$; for each of them (considered as H') the probability that (**) holds is least $t^{-1/3}$ and these events are independent for the various subsets. Let Φ_0 be the subset of those $\varphi \in \Phi$ which do not satisfy (3.6).

We estimate the probability of the following event B: $\exists \varphi \in \Phi_0$ $\varphi = \mathrm{range}(g)$ but there is no $\psi \in \Phi$ with $\psi \subseteq \mathrm{range}(f) \cup \varphi$, $\psi \cap \mathrm{range}(f) \neq \emptyset$, where f, g are taken independently from $\prod_H A$ and $\prod_G A$.

As we have seen for each fixed $\varphi \in \Phi_0$, the probability that B holds with this particular φ is at most $2^{-t^{1/3}}$, so if we first randomize g and then f we get $P(B) = p \leq 2^{-t^{1/3}}$. Consequently if we randomize first only f then with a probability of at least $1 - p^{1/2}$ we get an f so that if now we randomize g then B will hold with a probability of at most $p^{1/2}$. This implies that if the Lemma does not hold for some Φ, then it will also fail with $c \to c/2$ and $\Phi \to \Phi - \Phi_0$, therefore we may assume that (3.6) holds for all $\varphi \in \Phi$.

If our Lemma does not hold then, at least for some $f \in \prod_H A$, we have that the probability of (*) is greater than $t^{-l\beta}$. This implies in particular that if we take a random $g \in \prod_G A$ then the probability that $\mathrm{range}(g) \in \Phi$ is at least $t^{-l\beta}$. So if H' is a subset of H with $[|H|^{1/3}]$ elements and $f \in \prod_{H'} A$ is taken at random, then the expected number of sets $F \in H', |F| = l$ so that $f(F) \in \Phi$ is at least $\binom{|H'|}{l} t^{-\beta l} \geq (t^{1/4})^l t^{-\beta l} > 3$.

Now we compute this expected number E in a different way. For each $\varphi \in \Phi$ and $F \subseteq H'$, $|F| = l$ let $N_{\varphi,F}(f)$ be 1 if $\mathrm{range}(f_F) = \varphi$ and 0 otherwise. Clearly E is the expected value of $\sum_{\varphi \in \Phi, F \subseteq H'} N_{\varphi,F}$ that is $E = \sum_{\varphi,F} E(N_{\varphi,F})$. For each fixed φ, F the function $N_{\varphi,F}$ can be written in the form $K_{\varphi,F} + L_{\Phi,F}$, where $K_{\varphi,F}(f) = 1$ if $N_{\varphi,F} = 1$ and there is no $F' \neq F$, $F' \subseteq H'$, $|F'| = l$ with $\mathrm{range}(f|_{F'}) \in \Phi$; and $K_{\varphi,F}(f) = 0$ otherwise. (This implies that $L_{\varphi,F}(f) = 1$ if $N_{\varphi,F} = 1$ and there is an $F' \neq F$, $F' \subseteq H'$, $|F'| = l$ with $\mathrm{range}(f|_{F'}) \in \Phi$; and $L_{\varphi,F}(f) = 0$ otherwise.)

As we assumed each $\varphi \in \Phi$ satisfies condition (3.6). We claim that for each fixed φ and F we have $E(L_{\varphi,F}) \leq t^{-1/3} E(K_{\varphi,F})$. Indeed when we randomize f on $H' - F$ then (3.6) implies that the probability that there exists a $\psi \in \Phi$, $\psi \subseteq \mathrm{range}(f)$, $\mathrm{range}(f|_{H'-F}) \cap \psi \neq \emptyset$ is at most $t^{-1/3}$. That is $P(L_{\varphi,F} > 0) \leq t^{-1/3} P(\mathrm{range}(f|_F) = \varphi) \leq t^{-1/3} P(K_{\varphi,F} = 1)$. Therefore $E(N_{\varphi,F}) \leq 2E(K_{\varphi,F})$. The definition of $K_{\varphi,F}(f)$ implies that for a fixed f there can be only a single pair φ, F so that $K_{\varphi,F}(f) \neq 0$, that is the random variable $\sum_{\varphi,F} K_{\varphi,F}$ takes only the values 0 and 1 and therefore its expected value is at most 1. So we have $E \leq 2E(\sum_{\varphi,F} K_{\varphi,F}) \leq 2$ in contradiction to our previous lower bound.

Proof of Lemma 1.5. Suppose that $f \in \prod_{H \cup G} A$ so that (1.5) is not satisfied. We define two sequences of subsets of G: $G_1, ..., G_s, E_1, ..., E_s$ in the following way.

We define the pair G_i, E_i by induction on i. Our inductive definition will ensure that the sequences G_i, E_i meet the following requirements:

(3.7) $|G_i| \le t^\lambda$, $|E_i| \le t^\lambda$,

(3.8) all of the $2s$ sets G_i, E_i, $i = 1, ..., s$ are pairwise disjoint,

(3.9) none of the sets G_i is empty,

(3.10) $(1/4)t^{\mu-\lambda} \le |\bigcup_{i=1}^{s} G_i| \le (1/2)t^{\mu-\lambda}$.

For the proof we fix some arbitrary orderings of the sets $\lceil A \rceil^{t^\lambda} \times \Gamma$ and $\lceil A \rceil^{t^\lambda}$.

Assume that G_j, E_j have been already defined for all $j < i$. If $|\bigcup_{j=1}^{i} G_j| < (1/4)t^{\mu-\lambda}$ then (3.7) and (3.9) imply that $|\bigcup_{j<i} G_j \cup E_j| \le t^\mu$. Since 1.5 does not hold there is an element $\langle B, \gamma \rangle$ of $\lceil A \rceil^{t^\lambda} \times \Gamma$ so that

(3.11) $B \subseteq \mathrm{range}(f)$ and there is no $\varphi \subseteq \mathrm{range}(f|_{H \cup K_i})$, where $K_i = \bigcup_{j<i} G_j \cup E_j$ with $\varphi \in \Phi(B, \gamma)$.

Let B_i, γ_i be the smallest such pair and let $\varphi_i \in \Phi(B_i, \gamma_i)$, $\varphi_i \subseteq \mathrm{range}(f)$, so that $|\varphi_i - \mathrm{range}(f|_{H \cup K_i})|$ is minimal.

Suppose that $B_i = \mathrm{range}(f|_{E_i'})$, $\varphi_i = \mathrm{range}(f|_{G_i'})$ where E_i', G_i' are minimal sets with these properties. Moreover we also assume that among the minimal ones they are the first according to our fixed ordering of $\lceil A \rceil^{t^\lambda}$. Let

(3.12) $E_i = E_i' - (K_i \cup H)$, $G_i = G_i' - (K_i \cup H \cup E_i)$.

We check that requirements (3.7), (3.8), (3.9) are met.

(3.7) follows from $|\varphi_i| \le t^\lambda$, $|B_i| \le t^\lambda$. (3.12) implies (3.8), (3.11) implies (3.9).

We can continue defining the sequences G_i, E_i until we reach an i with $|\bigcup_{j<i} G_i| \ge (1/4)t^{\mu-\lambda}$. Let s be the first i with this property. (3.7) imply that (3.10) is satisfied too.

Let us suppose now that a positive integer $s \le t^\mu$ and a sequence $\bar{G}_1, ..., \bar{G}_s$, $\bar{G}_i \subseteq G$, $|\bar{G}_i| \le t^\lambda$ is fixed. Let $u = |\bar{G}_1 \cup ... \cup \bar{G}_s| = t^{\bar\mu}$. First we prove the following:

(**3.13**)   *The following statement implies Lemma 1.5:*

If $s, \bar{G}_1, ..., \bar{G}_s, u, \bar{\mu}$ are fixed (with the properties listed above) and f is taken at random from $\prod_{H \cup G}$ with uniform distribution, then the probability that (1.5) is not satisfied and the sequence G_i defined from f is identical to the sequence $\bar{G}_i$ is less than $t^{-\alpha t^{\bar{\mu}}}$, where α is an absolute constant.

Proof. 3.7, 3.10 and $G_i \subseteq G$, $|G| \leq t^\varepsilon$ implies that the number of possible sequences $\langle \bar{G}_i \rangle$ for a fixed u with the given properties is at most $t^{\varepsilon t^{\bar{\mu}}}$ and $\mu \geq \bar{\mu} \geq \mu - 2\lambda$. Therefore the assumption in (3.13) implies that the probability that a random f fails to satisfy (1.5) with a given $u = t^{\bar{\mu}}$ is at most $t^{\varepsilon t^{\bar{\mu}}} t^{-\alpha t^{\bar{\mu}}} \leq t^{-t^{\mu-3\lambda}}$. Since the number of possible integers u is at most t we have that the probability that f does not satisfy (1.5) is at most $t^{1-t^{\mu-3\lambda}} \leq t^{-\mu/2}$ which completes the proof of (3.13).

For the proof of the assumption in (3.13) let R be the set of all $f \in \prod_{H \cup G} A$ which do not satisfy condition (1.5) and if we define the sequence G_i from f then we get the sequence $\bar{G}_i$.

Assume that a sequence $\bar{G}_i$ is given and we randomize f first on the set $T = (H \cup G) - \bigcup_{i=1}^s \bar{G}_i$ and then successively on each $\bar{G}_i$, $i = 1, ..., s$. Assume that f has been already randomized on T and on $\mathcal{G}_i = \bigcup_{j<i} \bar{G}_j$. We will denote this function defined on $T \cup \mathcal{G}_i$ by f_i. Suppose that f_i has an extension $f \in R$. We show that

(**3.14**)   the function f_i uniquely determines the elements $B_j, \gamma_j, E_j, j \leq i$ belonging to any $f \in R$ with $f_i = f|_{T \cup \mathcal{G}_i}$.

Indeed we can prove the statement by induction on j. Suppose that (3.14) has been already proved for $1, ..., j-1$. B_j, γ_j is the smallest pair in $\lceil A \rceil^{t^\lambda} \times \Gamma$ which satisfies (3.11). (3.8) implies that $E'_j \subseteq T \cup \mathcal{G}_j$. So if we substitute the condition $B \subseteq \text{range}(f)$ by $B \subseteq \text{range}(f_j)$ in (3.11) then the definition gives the same pair B_j, γ_j and therefore E_j will depend only on f_i which completes the proof of (3.14). (Here we used the fact that according to the inductive hypothesis the sets $K_{...}$ appearing in (3.11) are uniquely determined by f_j).

We need the following statement for the conclusion of the proof:

(**3.15**)   *There exist absolute constants $c > 0$, $\alpha > 0$ so that if $\bar{G}_1, ..., \bar{G}_s$, $s \leq t^\mu$, $|\bar{G}_j| \leq t^\lambda$ for $j = 1, ..., s$ is fixed, $1 \leq i \leq s+1$ and h is picked at random with uniform distribution from $\prod_{T \cup \mathcal{G}_i} A$, then with a probability of at least $1 - 2^{-t^c}$ the following holds:*

if $h = f_i$ for some $f \in R$ and we take a random g from $\prod_{\bar{G}_i} A$ with uniform distribution, then the probability that there is a $\psi \in \Phi(B_i, \gamma_i)$ so that $\psi \subseteq$

range(f_i) $\cup$ range(g) but for any proper subset D of $\bar{G}_i$ there is no $\psi' \in$ $\Phi(B_i, \gamma_i)$ with $\psi' \subseteq$ range(f_i) $\cup$ range($g|_D$), is smaller than $t^{-\alpha|\bar{G}_i|}$.

First we show that (3.15) implies (3.13). Indeed if $f \in R$ then $g = f|_{\bar{G}_i}$ meets the requirement in (3.15) with $\psi = \varphi_i$. Therefore if f_i has an extension in R then with a probability of at least $1 - t^{-\alpha|\bar{G}_i|}$ the resulting function $f_i \cup g$ has no extension in R.

This implies that if we randomize f first on T and then successively on each G_i then either one of the functions f_i will be in the exceptional set for the first randomization in (3.15) (this happens with a probability of at most $s2^{-t^c} \le 2^{-t^{c/2}}$) or each f_i is outside the exceptional set for the first randomization. Suppose that this second condition, C_2 is satisfied. We estimate $P(f \in R|C_2)$. $P(f \in R$ and $C_2) \le t^{-\alpha(|\bar{G}_1|+\cdots+|\bar{G}_s|)}$ since when we randomize f on $\bar{G}_i$ then the probability that the resulting function still has an extension in R is at most $t^{-\alpha\bar{G}_i}$. $P(C_2) \ge 1 - 2^{-t^{c/2}}$, therefore $P(f \in R|C_2) \le 2t^{-\alpha(|\bar{G}_1|+\cdots+|\bar{G}_s|)}$ and so $P(f \in R) \le 2^{(-t^{c/2})} + (1 - 2^{-t^{c/2}})P(f \in R|C_2) \le 3t^{-\alpha(|\bar{G}_1|+\cdots+|\bar{G}_s|)} \le t^{-\alpha't^{\bar{\mu}}}$ which completes the proof of (3.13). Now we prove (3.15).

We estimate the probability of the event C: at the first randomization we get a function h which does not satisfy the properties given in the second paragraph of 3.15 and $h = f_i$ for some $f \in R$.

If $\gamma \in \Gamma$, $N \subseteq T \cup \mathcal{G}_i$, $|N| \le t^\lambda$ then we will denote by $C_{N,\gamma}$ the following event: C and $B_i =$ range($h|_N$) and $\gamma_i = \gamma$.

$P(C) \le \sum_{N,\gamma} P(C_{N,\gamma})$. Since the number of possible pairs N, γ is at most $t^{t^\lambda}2^{t^\lambda}$, it is sufficient to prove that for each fixed N and γ, $P(C_{N,\gamma}) \le 2^{-t^{c'}}$ for some absolute constant $c' > 0$.

Assume that N, γ are fixed. First we randomize the values of h on N. Our assumption implies that $B_i =$ range(h_N). Now we randomize h on the remaining points and g on $\bar{G}_i$. For the randomization of this pair we apply Lemma 3.1 with $H \to T \cup \mathcal{G}_i - N$, $G \to \bar{G}_i$, $\Phi \to \Phi(B_i, \gamma)$, $k \to t^\lambda$, $l \to |\bar{G}_i|$ and we get the required upper bound.

4. Proof of Theorem 1.2 and Theorem 1.3.

First we prove Theorem 1.3. For the proof of Theorem 1.2 we will actually need Theorem 1.3 in a somewhat stronger form:

Lemma 4.1 *Theorem 1.3 remains valid if we assume that an $a \in A$ is fixed and we replace (1.4) by*

$$\forall X_1 \in \langle\!\langle K - \{a\}\rangle\!\rangle_\Gamma^{t^\lambda} \exists X_2 \in \langle\!\langle K - \{a\}\rangle\!\rangle_\Gamma^{t^\lambda} ...QX_d \in \langle\!\langle K - \{a\}\rangle\!\rangle_\Gamma^{t^\lambda} \Psi(X_1, ..., X_d) \equiv$$

$$\equiv \exists U \in \lceil L - \{a\}\rceil^{t^\mu} \forall V \in \lceil L - \{a\}\rceil^{t^\mu} R_J(U, V) \equiv$$

$$\equiv \forall V \in \lceil L - \{a\}\rceil^{t^\mu} \exists U \in \lceil L - \{a\}\rceil^{t^\mu} R_J(U, V),$$

that is all of the quantifiers which were restricted to K resp. L are now restricted to $K - \{a\}$ resp. $L - \{a\}$.

We will prove the theorem in the original form. The proof of Lemma 4.1 is almost exactly the same with only trivial changes.

We prove Theorem 1.3 by induction on d. We start with $d = 2$. (The $d = 1$ case is a special case of $d = 2$.) The conclusion of the theorem for the $d = 2$ special case can be written in the following form:

$$\forall X \in \langle\!\langle K \rangle\!\rangle_\Gamma^{t^\lambda} \exists Y \in \langle\!\langle K \rangle\!\rangle_\Gamma^{t^\lambda} \Psi(X, Y) \equiv$$

$$\exists U \in \lceil L \rceil^{t^\mu} \forall V \in \lceil L \rceil^{t^\mu} R_J(U, V) \equiv \forall V \in \lceil L \rceil^{t^\mu} \exists U \in \lceil L \rceil^{t^\mu} R_J(U, V).$$

As we have indicated earlier we base the proof on Lemma 1.5. Following the idea given in the sketch of the proof we will define the relation R_Z for each $Z \subseteq A$ in the following way:

$$R_Z(U, V) \text{ iff } \forall X \in \langle\!\langle Z \cup V \rangle\!\rangle_\Gamma^{t^\lambda} \exists Y \in \langle\!\langle Z \cup U \cup X \rangle\!\rangle_\Gamma^{t^\lambda} \Psi(X, Y).$$

First we show that if the conditions of the theorem are satisfied then, with a probability of at least $1 - 2^{-t^c}$:

$$\forall X \in \langle\!\langle K \rangle\!\rangle_\Gamma^{t^\lambda} \exists Y \in \langle\!\langle K \rangle\!\rangle_\Gamma^{t^\lambda} \Psi(X, Y) \text{ implies } \exists U \in \lceil L \rceil^{t^\mu} \forall V \in \lceil L \rceil^{t^\mu} R_J(U, V).$$

We apply Lemma 1.5 with $\Phi(B, \gamma) = \{W \in \lceil A \rceil^{t^\lambda} | \exists \delta \in \Gamma \; \Psi(\langle B, \gamma \rangle, \langle W, \delta \rangle)\}$. According to Lemma 1.5 with a probability of at least $1 - 2^{-t^c}$ there exists a $T \subseteq L, |T| \le t^\mu$ so that

$$\forall \langle B, \gamma \rangle \in \langle\!\langle K \rangle\!\rangle_\Gamma^{t^\lambda} \exists W \subseteq T \cup J \cup B \; W \in \Phi(B, \gamma),$$

that is $\exists U \in \lceil L \rceil^{t^\mu} \forall V \in \lceil L \rceil^{t^\mu} R_J(U, V)$ holds with $U = T$, that is

$$\forall X \in \langle\!\langle K \rangle\!\rangle_\Gamma^{t^\lambda} \exists Y \in \langle\!\langle K \rangle\!\rangle_\Gamma^{t^\lambda} \Psi(X, Y) \to \exists U \in \lceil L \rceil^{t^\mu} \forall V \in \lceil L \rceil^{t^\mu} R_J(U, V).$$

The definition of R_Z implies that for any choice of K we have:

$$\exists U \in \lceil L \rceil^{t^\mu} \forall V \in \lceil L \rceil^{t^\mu} R_J(U, V) \to \forall V \in \lceil L \rceil^{t^\mu} \exists U \in \lceil L \rceil^{t^\mu} R_J(U, V) \to$$

$$\forall X \in \langle\!\langle K \rangle\!\rangle_\Gamma^{t^\lambda} \exists Y \in \langle\!\langle K \rangle\!\rangle_\Gamma^{t^\lambda} \Psi(X, Y)$$

which completes the proof of the $d = 2$ case.

Now we show that the $d = 2$ case holds under slightly more general circumstances, namely it remains true for formulas of the type

$$(4.1) \quad \forall X_1 \in \langle\!\langle K \rangle\!\rangle_\Gamma^{t^\lambda} ... \forall X_i \in \langle\!\langle K \rangle\!\rangle_\Gamma^{t^\lambda} \exists Y_1 \in \langle\!\langle K \rangle\!\rangle_\Gamma^{t^\lambda} ... \exists Y_i \in \langle\!\langle K \rangle\!\rangle_\Gamma^{t^\lambda}$$

$$\Psi(X_1, ..., X_i, Y_1, ..., Y_i),$$

where i is an arbitrary constant. More precisely the following holds:

Lemma 4.2 *For each fixed positive integer i the assertion of Theorem 1.3 in the $d = 2$ case remains valid, if we assume that Ψ is a $2i$-ary relation on $\langle\!\langle A \rangle\!\rangle_\Gamma^{t^\lambda}$ and we replace the first formula of (1.4) by (4.1).*

Since we have already proved the $d = 2$ case it is sufficient to show that the formula in (4.1) is equivalent to a formula of the same type with $i = 1$. We will show that such a formula indeed exists although for different λ and Γ. More precisely we prove the following:

Suppose that c, λ, Ψ are as in the modified version of Theorem 1.3, $\lambda < c/2$. Then there exists a set Γ', $|\Gamma'| \leq 2^{t^{2c}}$ and a binary relation Ψ' on $\langle\!\langle A \rangle\!\rangle_{\Gamma'}^{t^\lambda}$ so that if f is an arbitrary element of $\prod_{H \cup G} A$ and range$(f) = K$, then

$$\forall X_1 \in \langle\!\langle K \rangle\!\rangle_\Gamma^{t^\lambda} ... \forall X_i \in \langle\!\langle K \rangle\!\rangle_\Gamma^{t^\lambda} \exists Y_1 \in \langle\!\langle K \rangle\!\rangle_\Gamma^{t^\lambda} ... \exists Y_i \in \langle\!\langle K \rangle\!\rangle_\Gamma^{t^\lambda}$$

$$\Psi(X_1, ..., X_i, Y_1, ..., Y_i) \equiv \forall X \in \langle\!\langle K \rangle\!\rangle_{\Gamma'}^{t^{i\lambda}} \exists Y \in \langle\!\langle K \rangle\!\rangle_{\Gamma'}^{t^{i\lambda}} \Psi'(X, Y).$$

We will code each sequence $X_1, ..., X_i$, $X_j \in \langle\!\langle A \rangle\!\rangle_\Gamma^{t^\lambda}$ by a single $W \in \langle\!\langle A \rangle\!\rangle_{\Gamma'}^{t^{i\lambda}}$. Moreover this coding will have the additional property that if $X_1 = \langle Z_1, \gamma_1 \rangle, ...$ then $W = \langle Z_1 \cup ... \cup Z_i, \delta \rangle$, and therefore if $X_j \in \langle\!\langle K \rangle\!\rangle_\Gamma^{t^\lambda}$, $j = 1, ..., i$ then $W \in \langle\!\langle K \rangle\!\rangle_{\Gamma'}^{t^{i\lambda}}$.

Let Γ' be an arbitrary set with $2^{t^{ic}}$ elements. We want to define a map ι of $\langle\!\langle A \rangle\!\rangle_{\Gamma'}^{t^{i\lambda}}$ onto $P = \prod_{j=1}^i \langle\!\langle A \rangle\!\rangle_\Gamma^{t^\lambda}$. (A $p \in P$ will be coded by a suitable element of $\iota^{-1}(p)$). First for each fixed $X \in \lceil A \rceil^{t^{i\lambda}}$ let ι_X be a map defined on $\{X\} \times \Gamma'$ so that if $p = \langle ..., \langle X_j, \gamma_j \rangle, ... \rangle \in P$ and $\forall j \ X_j \subseteq X$ then there is a $\gamma' \in \Gamma'$ with $\iota_X(\langle X, \gamma' \rangle) = p$. Since the number of possible sequences p is at most $(t^{t^\lambda} 2^{t^c})^i \leq 2^{t^{2c}} = |\Gamma'|$ such a map ι_X does exist.

Let ι be the common extension of all of the maps ι_X. Now we can define Ψ' by $\Psi'(\langle X, Y \rangle)$ iff $\Psi(\iota(X) \circ \iota(Y))$ where $\iota(X) \circ \iota(Y)$ is the concatenation of the sequences $\iota(X), \iota(Y)$. Q.E.D.

In the same way as we have proved Lemma 4.2 we may show that Lemma 1.5 also holds in a slightly more general form. We will need this form of Lemma 1.5 in the proof of Theorem 1.2

Lemma 4.3 *For all positive integers i and sufficiently small $\varepsilon > 0$ if $\mu > 0$ is sufficiently small with respect to ε and $\lambda > 0$ is sufficiently small with respect to μ, t is sufficiently large, A, H, G, Γ are finite sets, $H \cap G = \emptyset$, $|G| = [t^\varepsilon]$, $|H| + |G| = t$, $|\Gamma| \leq 2^{t^\lambda}$ and for all $B_1, ..., B_i \in [A]^{t^\lambda}$ and $\gamma \in \Gamma$, $\Phi(B_1, ..., B_i, \gamma)$ is a subset of the powerset of $[A]^{t^\lambda}$, then the following holds:*

Suppose that f is taken at random with uniform distribution from $\prod_{H \cup G} A$. Then with a probability of at least $1 - 2^{-t^\mu}$ we have that:

If for all $B_1, ..., B_i \in \lceil \text{range}(f) \rceil^{t^\lambda}$, and $\gamma \in \Gamma$ there is a $\varphi \subseteq \text{range}(f)$ so that $\varphi \in \Phi(B_1, ..., B_i, \gamma)$, then there is a subset M of G with at most t^μ elements so that for each $B_1, ..., B_i \in \lceil \text{range}(f) \rceil^{t^\lambda}, \gamma \in \Gamma$ there is a $\varphi \subseteq \text{range}(f|_{H \cup M}) \cup B_1 \cup ... \cup B_i$ with $\varphi \in \Phi(B_1, ..., B_i, \gamma)$.

The following Lemma will make it possible to replace quantification over $\langle\!\langle K \rangle\!\rangle_\Gamma^{t^\lambda}$ by quantification over $\langle\!\langle L \rangle\!\rangle_{\Gamma'}^{t^\lambda}$ for a suitably chosen but still not too large Γ'.

Lemma 4.4 *If $\lambda > 0$ is sufficiently small and t is a sufficiently large integer then the following holds:*

Suppose that A is a finite set. Then there exists a set Γ' with at most $2^{t^{2\lambda}}$ elements so that for all $J \subseteq [A]^t$ there is a function ν_J defined on $[A]^{t^\lambda} \times \Gamma'$ with the following properties:

(a) $\text{range}(\nu_J) \subseteq [A]^{t^\lambda}$

(b)   *for each $S \in [A]^{t^\lambda}$ there is a $\gamma' \in \Gamma'$ so that $\nu_J(S - J, \gamma') = S$.*

The lemma guarantees that for a fixed J we may code the elements of $[A]^{t^\lambda}$ by pairs X, γ' where $X \in [A - J]^{t^\lambda}$ and $\gamma \in \Gamma'$.

Proof. Assume that $J \in [A]^t$ and an $X \in [A - J]^{t^\lambda}$ are fixed. The number of sets $S \in [A]^{t^\lambda}$ with $S - J = X$ is at most $t^{t^\lambda} \leq 2^{t^{2\lambda}}$. Therefore there exists a map $\eta_J^{(X)}$ of Γ' onto $\{S \in [A]^{t^\lambda} | S - J = X\}$. We may define ν_J by $\nu_J(X, \gamma') = \eta_J^{(X)}(\gamma')$. Q.E.D.

We will have frequently the following situation. We will be able to prove that any fixed $S \in [A]^{t^\lambda}$ with high probability (for the randomization of K) satisfies a property Φ. Under certain circumstances we will be able to conclude from this that with high probability all of the elements of $[K]^{t^\lambda}$ have property Φ. Namely if we want to prove that $\forall S \in [K]^{t^\lambda} \Phi(S)$, then it will be enough to show that for each fixed $U \in [H \cup G]^{t^\lambda}$ the probability of $\forall S \in \lceil K \rceil^{t^\lambda} S = \text{range}(f|_U) \to \Phi(S)$ is close to one. The following trivial Lemma describes this situation under slightly more general circumstances.

Lemma 4.5 *If $\lambda > 0$, and t is a positive integer then the following holds:*

Suppose that $0 < q < 1$, Γ, A, H, G are finite sets $|G \cup H| = t$, $G \cap H = \emptyset$, $|\Gamma| \leq 2^{t^\lambda}$ and Θ is a unary relation on the set $2^A \times 2^A \times \langle\!\langle A \rangle\!\rangle_\Gamma^{t^\lambda} \times 2^{H \cup G}$.

Assume that f is a random element of $\prod_{H \cup G} A$ with uniform distribution, $\text{range}(f) = K$, $\text{range}(f|_H) = J$ and $\text{range}(f|_G) = L$.

If for each fixed $x = \langle S, \gamma \rangle \in \langle\!\langle A \rangle\!\rangle_\Gamma^{t^\lambda}$ and $U \in \lceil H \cup G \rceil^{t^\lambda}$ we have $P(\Theta(J, L, x, U) \mid S$ $\text{range}(f|_U)) \geq 1 - q$, then

$$P(\forall y = \langle S, \gamma \rangle \in \langle\!\langle K \rangle\!\rangle_\Gamma^{t^\lambda}, \forall U \in \lceil H \cup G \rceil^{t^\lambda}, \ S = \text{range}(f|_U) \to \Theta(J, L, y, U)) \geq$$
$$1 - t^{2t^\lambda} q.$$

Proof. For each $y \in \langle\!\langle A \rangle\!\rangle_\Gamma^{t^\lambda}$, $U \in \lceil H \cup G \rceil^{t^\lambda}$ let $M_{y,U}$ be the event $S = \text{range}(f|_U) \to \Theta(J, L, y, U)$.

$\neg \forall y, U \ M_{y,U}$ implies that $\exists U \in \lceil G \cup H \rceil^{t^\lambda}$, $S \in \lceil K \rceil^{t^\lambda}$ and $\gamma \in \Gamma$ so that $\neg M_{\langle S,\gamma \rangle, U}$ and $S = \text{range}(f|_U)$. Therefore $P(\neg \forall \ y \in \langle\!\langle K \rangle\!\rangle_\Gamma^{t^\lambda} \forall U \in \lceil H \cup G \rceil^{t^\lambda} \ M_{y,U}) \leq \sum_{U,\gamma} \sum_S P(S = \text{range}(f|_U)) P(\neg M_{\langle S,\gamma \rangle, U} \mid S = \text{range}(f|_U))$. Since $|\lceil G \cup H \rceil^{t^\lambda}| \leq t^{t^\lambda}$ and $|\Gamma| \leq 2^{t^\lambda}$ the number of choices for the pair $\langle U, \gamma \rangle$ is at most $t^{t^\lambda} 2^{t^\lambda} \leq t^{2t^\lambda}$. Thus using the upper bounds $P(\neg M_{\langle S,\gamma \rangle, U} \mid S = \text{range}(f|_U)) \leq q$ and $\sum_S P(S = \text{range}(f|_U)) \leq 1$ we get the assertion of the Lemma. Q.E.D.

Now we complete the inductive proof of Theorem 1.3. We get an equivalent statement if we replace the first formula of (1.4) by

$$(4.2) \quad \exists X_1 \in \langle\!\langle K \rangle\!\rangle_\Gamma^{t^\lambda} \forall X_2 \in \langle\!\langle K \rangle\!\rangle_\Gamma^{t^\lambda} \ldots \mathcal{Q} X_d \in \langle\!\langle K \rangle\!\rangle_\Gamma^{t^\lambda} \ \Psi(X_1, \ldots, X_d)$$

(where $\mathcal{Q}$ is the quantifier $\forall$ if d is odd and $\mathcal{Q}$ is the quantifier $\exists$ if d is even) since the conclusion of the theorem is symmetric to negation. This equivalence clearly holds for every fixed d so we may prove the theorem in the new form while using the old one in the inductive assumption.

Suppose now that $d > 2$ and the theorem holds for $d \to d - 1$. The following Lemma shows that under this assumption if we fix J at random then with high probability any formula of the type (4.2) is equivalent to a Σ_2 formula in the sense of Theorem 1.3. (Note that this Σ_2 formula is of somewhat more complicated type than the one given in Theorem 1.3, so we will have to use Lemma 4.2 to conclude the proof.)

Lemma 4.6 *For all integers $d > 2$ if Theorem 1.3 holds for $d \to d - 1$ then there exists $\varepsilon > 0$ so that for all $\mu > 0$ if $c > 0$ is sufficiently small, $\lambda > 0$ is sufficiently small with respect to c and t is sufficiently large then the following holds:*

Suppose that Γ, A, H, G are finite sets $|\Gamma| \le 2^{t^c}$, $|G \cup H| = t$, $G \cap H = \emptyset$, $|G| \le t^\varepsilon$, Ψ is a d-ary relation on $\langle\!\langle A \rangle\!\rangle_\Gamma^{t^\lambda}$. Then for each $Z \subseteq A$ there exists a set Γ' with at most $2^{t^{2c}}$ elements and a unary relation R_Z on $\langle\!\langle A \rangle\!\rangle_{\Gamma'}^{t^\lambda} \times \lceil A \rceil^{t^\mu} \times \lceil A \rceil^{t^\mu}$ so that if f is a random element of $\prod_{H \cup G} A$ with uniform distribution, $\mathrm{range}(f) = K$, $\mathrm{range}(f|_H) = J$ and $\mathrm{range}(f|_G) = L$, then with a probability of at least $1 - 2^{-t^c}$ the following holds:

$$(4.3) \quad \exists X_1 \in \langle\!\langle K \rangle\!\rangle_\Gamma^{t^\lambda} \forall X_2 \in \langle\!\langle K \rangle\!\rangle_\Gamma^{t^\lambda} \dots \mathcal{Q} X_d \in \langle\!\langle K \rangle\!\rangle_\Gamma^{t^\lambda}$$

$$\Psi(X_1, ..., X_d) \equiv \exists Z \in \langle\!\langle L \rangle\!\rangle_{\Gamma'}^{t^\lambda} \exists U \in \lceil L \rceil^{t^\mu} \forall V \in \lceil L \rceil^{t^\mu} R_J(Z, U, V).$$

Proof. We may write the first formula of (4.3) in the form of $\exists X \in \langle\!\langle K \rangle\!\rangle_\Gamma^{t^\lambda} \Phi(X)$ where Φ contains only $d - 1$ quantifiers. We apply the inductive assumption, that is Theorem 1.3 in the $d \to d - 1$ case, for each formula $\Phi(X)$, where $X \in \langle\!\langle A \rangle\!\rangle_\Gamma^{t^\lambda}$ is fixed.

We will show that if we take a random f then with high probability the conclusion of the theorem holds simultaneously for all $\Phi(X)$, $X \in \langle\!\langle K \rangle\!\rangle_\Gamma^{t^\lambda}$. We will use Lemma 4.5 to prove this fact. We define the relation Θ from the Lemma in the following way:

$\Theta(J, L, \langle S, \gamma \rangle, W)$ iff "$J, L \subseteq A$ and if $|W| = |S|$, $|W - H| = |S - J|$ then there exists a binary relation $R_{J,W}^y$, (where $y = \langle S, \gamma \rangle$) on $\lceil A \rceil^{t^\mu}$ so that

$$(4.4) \quad \Phi(y) \equiv$$
$$\exists U \in \lceil L \rceil^{t^\mu} \forall V \in \lceil L \rceil^{t^\mu} R_{J,W}^y(U, V) \equiv \forall V \in \lceil L \rceil^{t^\mu} \exists U \in \lceil L \rceil^{t^\mu} R_{J,W}^y(U, V)."$$

The relation $R_{J,W}^y$ formally depends on W. However since W does not occur anywhere else in condition (4.4) we may suppose that there is a single relation $R_J^y = R_{J,W}^y$ for all possible W.

Assume that $y = \langle S, \gamma \rangle \in \langle\!\langle A \rangle\!\rangle_\Gamma^{t^\lambda}$ and $W \subseteq H \cup G$, $|W| = |S|$ and a function $\bar{f}$ mapping W onto S is fixed. We apply the inductive assumption with $H \to H - W$, $G \to G - W$ $f \to \tilde{f}$, $J \to \tilde{J}$, $L \to \tilde{L}$, $K \to \tilde{K}$. The conclusion implies that if $f = \bar{f} \cup \tilde{f}$, $J = \tilde{J} \cup \mathrm{range}(f|_{W-G})$, $L = \tilde{L} \cup \mathrm{range}(f|_{W-H})$, $K = J \cup L$, then with a probability of at least $1 - 2^{-t^c d - 1}$ we have $\Theta(J, L, y, W)$. Moreover as we have seen we may suppose that the relation $R_{J,W}^y$ does not depend on W.

Therefore according to Lemma 4.5 with a probability of at least $1-t^{2t^\lambda}2^{-t^{c_{d-1}}} \geq 1 - 2^{t^{4c}}$ we have that

$$(4.5) \quad \Phi(y) \equiv$$
$$\exists U \in \lceil L \rceil^{t^\mu} \forall V \in \lceil L \rceil^{t^\mu} R_J^y(U,V) \equiv \forall V \in \lceil L \rceil^{t^\mu} \exists U \in \lceil L \rceil^{t^\mu} R_J^y(U,V)$$

holds simultaneously for all $y \in \langle\!\langle K \rangle\!\rangle_\Gamma^{t^\lambda}$. (Here we used that we picked c sufficiently small with respect to c_{d-1} the number corresponding to c in the inductive assumption.)

Assume now that we randomize f only on H. According to the previous remark with a probability of at least $1 - 2^{-t^{2c}}$ we get a set J with the following property:

if we continue the randomization of f then with a probability of at least $1 - 2^{-t^{2c}}$ we get a set L so that (4.5) holds for all $y \in \langle\!\langle K \rangle\!\rangle_\Gamma^{t^\lambda}$. Assume that a J is fixed with this property.

If we continue the randomization then with a probability of at least $1 - 2^{t^{2c}}$, our original formula $\exists X \in \langle\!\langle K \rangle\!\rangle_\Gamma^{t^\lambda} \Phi(X)$ holds for some K iff

$$(4.6) \quad \exists X \in \langle\!\langle K \rangle\!\rangle_\Gamma^{t^\lambda} \exists U \in \lceil L \rceil^{t^\mu} \forall V \in \lceil L \rceil^{t^\mu} R_J^X(U,V).$$

Since J is fixed and $K = J \cup L$, Lemma 4.4 (with $W \to J$) implies that there exists a Γ' with at most $2^{t^{2c}}$ elements so that (4.6) is equivalent to the following formula:

$$(4.7) \quad \exists Z \in \langle\!\langle L \rangle\!\rangle_{\Gamma'}^{t^\lambda} \exists U \in \lceil L \rceil^{t^\mu} \forall V \in \lceil L \rceil^{t^\mu} R_J^{\nu_J(Z)}(U,V)$$

where ν_J is a suitable function depending only on J, which completes the proof of Lemma 4.6.

Lemma 4.7 *There exists $\varepsilon > 0$ so that for all $\mu > 0$ if $c > 0$ is sufficiently small with respect to μ, $\lambda > 0$ is sufficiently small with respect to c and t is sufficiently large then the following holds:*

Suppose that Γ, A, H, G are finite sets, $|G \cup H| = t$, $G \cap H = \emptyset$, $|G| \leq t^\varepsilon$, Ψ is a unary relation on $\langle\!\langle A \rangle\!\rangle_\Gamma^{t^\lambda} \times \lceil A \rceil^{t^\lambda} \times \lceil A \rceil^{t^\lambda}$. Then for each $Z \subseteq A$ there exists a binary relation R_Z on $\lceil A \rceil^{t^\mu}$ so that if f is a random element of $\prod_{H \cup G} A$ with uniform distribution, $\text{range}(f) = K$, $\text{range}(f|_H) = J$ and $\text{range}(f|_G) = L$, then with a probability of at least $1 - 2^{-t^c}$ the following holds:

$$(4.8)$$

$$\exists Z \in \langle\!\langle K \rangle\!\rangle_\Gamma^{t^\lambda} \exists U \in \lceil K \rceil^{t^\lambda} \forall V \in \lceil K \rceil^{t^\lambda} \Psi(Z,U,V) \equiv$$
$$\exists U \in \lceil L \rceil^{t^\lambda} \forall V \in \lceil L \rceil^{t^\lambda} R_J(U,V) \equiv \forall U \in \lceil L \rceil^{t^\lambda} \exists V \in \lceil L \rceil^{t^\lambda} R_J(U,V) .$$

Proof. The statement of the Lemma is an immediate consequence of the already proven $d = 2$ case of the theorem and Lemma 4.2.

Now we may finish easily the proof of Theorem 1.3. Indeed let $\varepsilon' > 0$ is sufficiently small and $\varepsilon > 0$ is sufficiently small with respect to ε' and suppose that $G \subseteq G' \subseteq G \cup H$, $|G'| = [t^{\varepsilon'}]$, $H' = (G \cup H) - G'$. First we apply Lemma 4.6 with $\varepsilon \to \varepsilon'$, $G \to G'$, $H \to H'$. Let $J' = \mathrm{range}(f|H')$. We have that with a probability of at least $1 - 2^{t^{c/2}}$ we get a J' so that if now we randomize L' then with a probability of at least $1 - 2^{t^{c/2}}$ (4.3) holds with $J \to J'$, $L \to L'$. Suppose that such a J' is fixed. Now we apply Lemma 4.7 with $G \to G$, $H \to G' - G$, $t \to |G'|$. The conclusion of the Lemma is the assertion required by Theorem 1.3.

Proof of Theorem 1.2. First we show that it is enough to prove the theorem for a special type of Σ_2 formula.

Lemma 4.8 *Assume that Theorem 1.2 holds if $d = 2$ and the formula in (1.3) is of the form*

$$\exists X \subseteq \lceil K - \{a\}\rceil^{t^{\varepsilon}} \forall Y \subseteq \lceil K - \{a\}\rceil^{t^{\varepsilon}} \Psi_a(X, Y).$$

Then Theorem 1.2 holds in the general case too.

Let $\Theta_a(K)$ be the formula in (1.3). According to Lemma 4.1 (the stronger version of Theorem 1.3) for any fixed $a \in A$ and $Z \subseteq A$ there exists a binary relation R_Z^a on $\lceil A\rceil^{t^{u}}$ so that if we take J, L, K at random as described in the theorem, then with a probability of at least $1 - 2^{t^c}$ we have:

$$(4.9) \quad \Theta_a(K) \equiv \exists U \in \lceil L - \{a\}\rceil^{t^{u}} \forall V \in \lceil L - \{a\}\rceil^{t^{u}} R_J^a(U, V) \equiv$$

$$\forall V \in \lceil L - \{a\}\rceil^{t^{u}} \exists U \in \lceil L - \{a\}\rceil^{t^{u}} R_J^a(U, V).$$

Using Lemma 4.5 we show that with a probability of at least $1 - 2^{-t^{c/2}}$, (4.9) holds simultaneously for all $a \in K$. Indeed if we define the relation Θ of Lemma 4.5 by $\Theta(J, L, \langle S, \gamma\rangle, U)$ iff "if $|U| = |S| = 1$ then (4.9) holds" then the conclusion of the Lemma and $\lambda < c$ implies our assertion.

We will also need an upper bound on the number of elements $a \in A$ (not necessarily in K) so that (4.9) does not hold. Since for each fixed $a \in A$ it holds with a probability of at least $1 - 2^{-t^c}$, the expected number of elements a where it does not hold is at most $2^{-t^c}|A|$, therefore

(4.10) with a probability of at least $1 - 2^{-t^{c/2}}$ the number of elements where (4.9) does not hold is at most $2^{-t^{c/2}}|A|$.

Assume now that we have performed the randomization of J. With a probability of at least $1 - 2^{-t^{c/4}}$ we get a J so that for the remaining part of the randomizaton (4.9) holds simultaneously for each $a \in K$ with a probability of at least $1 - 2^{-t^{c/4}}$. Suppose that such a J is fixed. (4.10) implies that we may assume that J has the following additional property:

(4.11)   if we randomize L then with a probability of at least $1 - 2^{-t^{c/4}}$ the number of elements where (4.9) does not hold is at most $2^{-t^{c/4}} |A|$.

These remarks imply that we have to prove the original statement of theorem with the following changes: we randomize L instead of K and the formula in (1.3) is substituted by

(4.12) $\exists U \in \lceil L - \{a\} \rceil^{t^\mu} \forall V \in \lceil L - \{a\} \rceil^{t^\mu} R_J^a(U, V)$.

Indeed we have seen that with high probability the two formulae are equivalent for each $a \in K$, and (4.11) implies that with high probability if the original set S has at least $t^{-\varepsilon} |A|$ elements, the new "S" defined by "the set of all $a \in A$ where (4.12) holds" has at least $t^{-2\varepsilon} |A|$ elements.

The assumptions of Lemma 4.8 however guarantee that the theorem holds in this case. Q.E.D.

Now we prove the theorem in the special case described in Lemma 4.8. Let $\varepsilon' \gg \tau \gg \mu \gg \varepsilon > 0$, where $\alpha \gg \beta$ means that β is sufficiently small with respect to α.

We have to prove that with high probability either the density of S is small, or

$$\exists a \in K \exists X \subseteq \lceil K - \{a\} \rceil^{t^\varepsilon} \forall Y \subseteq \lceil K - \{a\} \rceil^{t^\varepsilon} \Psi_a(X, Y).$$

Let $G \cup H = A$, $G \cap H = \emptyset$, $|G| = [t^{\varepsilon'}]$. According to Lemma 4.3 (the stronger form of Lemma 1.5), with high probability for the randomization of K, this either holds or

(4.13) $\exists U \in \lceil L \rceil^{t^\mu} \forall a \in K, X \in \lceil K - \{a\} \rceil^{t^\varepsilon} \exists Y \subseteq \lceil J \cup U - \{a\} \rceil^{t^\varepsilon} \neg \Psi(X, Y)$.

We will show that for every fixed $U \in \lceil A \rceil^{t^\mu}$,

(4.14)   the probability that (4.13) holds and $|S|/|A| \geq t^{-\varepsilon}$ is smaller than 2^{-t^τ}.

This fact and Lemma 4.5 implies that that the probability that (4.13) holds for each $U \in \lceil K \rceil^{t^\mu}$ and $|S|/|A| \geq t^{-\varepsilon}$ is smaller than $2 - t^{\tau/2}$ which will complete the proof of the Theorem. Therefore the only missing part of the proof of Theorem 1.2 is (4.14) which follows from Lemma 2.1, stated at the and of Section 2.

Proof of Lemma 2.1. Let $\lambda > 0$ be sufficiently small and $\xi > 0$ be sufficiently small with respect to λ. Suppose that an arbitrary ordering is fixed on the power set of F, $f \in \prod_F A$, $|\bar{S}_f| \leq t^{-\xi}|A|$, $|\bar{S}_f \cap \text{range}(f)| < t^\xi$ and for each $a \in A$ if there is a $Y \subseteq F$, $|Y| \leq t^\xi$ and $\text{range}(f|_Y) \in \Phi_a$ then Y_a is the least Y with this property. For each $y \in F$ let $D_y = \{a \in A \mid y \in Y_a\}$ and $d_y = |D_y||A|^{-1}$. Clearly $\sum_{y \in F} d_y = \sum_{a \in A} F \sum_{y \in Y_a} |A|^{-1} \leq t^\xi$. Let $M = \{y \in F \mid d_y \geq t^{-1/2}\}$. Since $\sum_{y \in F} d_y \leq t^\xi$ we have $|M| \leq t^{\xi+1/2}$.

Let Z be a random subset of F with $[t^{1/4}]$ elements. Clearly with a probability of at least $1 - 2^{-t^\lambda}$ we have that $|Z \cap M| \leq t^{10\lambda}$. Let $Z \cap M = N_f$.

We prove that for every subset N of F with at most $t^{10\lambda}$ elements if we take both $f \in \prod_F A$ and $Z \subseteq F$, $|Z| = [t^{1/4}]$ at random, independently and with uniform distribution, then the probability of the following event is at most $t^{t^{-11\lambda}}$:

(4.15) $\quad |\bar{S}_f| \geq t^{-\xi}|A|$, $|\bar{S}_f \cap \text{range}(f)| < t^\xi$ and $N_f = N$.

Since there are at most $t^{t^{10\lambda}}$ possible choices for the set N we have that $P(|\bar{S}_f| \leq t^{-\xi}|A|$ and $|\bar{S}_f \cap \text{range}(f)| < t^\xi) \leq t^{t^{-\lambda}}$ which implies the statement of the Lemma.

Now we prove that the probability of (4.15) is at most $t^{t^{-11\lambda}}$. Assume that N is fixed. First we randomize Z and $f|_{(F-Z) \cup N}$. Let W be a subset of $Z - N$ with $[t^{20\lambda}]$ elements. We randomize f on all of the remaining points of $F - W$. We claim that independently of the result of all of the randomizations the probability of (4.15) is at most $t^{-t^{11\lambda}}$. Since $N \cap W = \emptyset$ for each $w \in W$ we have that $|\{a \in A \mid w \in Y_a\}| \leq t^{-1/2}|A|$. Therefore if we define S' by

$$S' = \{a \in A \mid \exists U \in \Phi_a \quad U \subseteq \text{range}(f|_{F-W})\},$$

then we have $|S'| \geq |\bar{S}_f| - t^{20\lambda}t^{-1/2}|A|$. So if $|S'| \leq (1/2)t^{-\xi}|A|$ then (4.15) cannot hold. Suppose that $|S'| > (1/2)t^{-\xi}|A|$. Now we randomize f on W. Since for each fixed $w \in W$ the probability of $f(w) \in S'$ is at least $(1/2)t^{-\xi}$, $|W| \geq t^{20\lambda}$ and $\xi > 0$ is sufficiently small with respect to λ, Chernoff's inequality implies that with a probability of at least $1 - 2^{-t^{19\lambda}}$ there will be at least t^ξ elements $w \in W$ with $f(w) \in S' \subseteq \bar{S}_f$. Q.E.D.

References

[Aj] Ajtai M, Σ_1^1-formulae on finite structures. Annals of Pure and Applied Logic 24 (1983), 1-48.

[Aj1] Ajtai M, *First order definability on finite structures*. Annals of Pure and Applied Logic 45 (1989) 211-225.

[AW] Ajtai M. and Wigderson A., *Deterministic simulation of probabilistic constant depth circuits*. Advances in Computing Research, Volume 5, 199-222.

[FSS] Furst M., Saxe J.B and Sipser M., *Parity, Circuits and the Polynomial Time Hierarchy*. Mathematical Systems Theory 17 (1984), 13-17.

[H] Håstad J., *Almost optimal lower bounds for small depth circuits*. Proc. 18th Ann. ACM Symp. Theory Computing (1986) 6-20.

[Y] Yao A., *Separating the polynomial time hierarchy by oracles*. Proc. 26th Ann. IEEE Symp. Found. Comp. Sci. (1985) 1-10.

Networks Computing Boolean Functions for Multiple Input Values

Dietmar Uhlig [*]

Abstract

Let f be an arbitrary Boolean function depending on n variables and let A be a network computing them, i.e., A has n inputs and one output and for an arbitrary Boolean vector a of length n outputs $f(a)$. Assume we have to compute simultaneously the values $f(a^1), ..., f(a^r)$ of f on r arbitrary Boolean vectors $a^1, ..., a^r$. Then we can do it by r copies of A. But in most cases it can be done more efficiently (with a smaller complexity) by one network with nr inputs and r outputs (as already shown in Uhlig (1974)). In this paper we present a new and simple proof of this fact based on a new construction method. Furthermore, we show that the depth of our network is "almost" minimal.

1. Introduction

Let us consider (combinatorial) networks. Precise definitions are given in [Lu58, Lu65, Sa76, We87]. We assume that a complete set G of gates is given, i.e., every Boolean function can be computed (realized) by a network consisting of gates of G. For example, the set consisting of 2-input AND, 2-input OR and the NOT function is complete. A cost $C(G_i)$ (a positive number) is associated with each of the gates $G_i \in G$. The complexity $C(A)$ of a network A is the sum of the costs of its gates. The complexity $C(f)$ of a Boolean function f is defined by $C(f) = \min C(A)$ where A ranges over all networks computing f.

By B_n we denote the set of Boolean functions $\{0,1\}^n \longrightarrow \{0,1\}$. We define $C(n) = \max_{f \in B_n} C(f)$. The function $C(n)$ was first investigated by C. Shannon [Sh49] in his pioneering paper. Hence it is called the *Shannon function.*

---

[*]Ingenieurhochschule Mittweida, Technikumplatz 17, Mittweida, O-9250, Germany

Lupanov showed in 1958 in [Lu58, Lu65] that

$$C(n) \sim \frac{\rho \cdot 2^n}{n} \qquad (1)$$

where $\rho = \rho(G) = $ const. ($a(n) \sim b(n)$ means that there is a function $\alpha(n) \longrightarrow 0$ as $n \longrightarrow \infty$ such that $a(n) = b(n)(1 + \alpha(n))$.) The constant ρ in (1) will be determined by

$$\rho = \rho(G) = \min \frac{C(G_i)}{n_i - 1} \qquad (2)$$

where G_i ranges over all elements of G with an input number $n_i > 1$. A gate for which (2) holds is called a *cheapest* gate.

It is well known that the number of elements of B_n is equal to 2^{2^n}. Let $N_\epsilon(n)$ be the number of functions $f \in B_n$ for which $C(f) \leq (1 - \epsilon)C(n)$. Let $N_0(n) = N(n)$. Obviously, $N(n) = 2^{2^n}$. Lupanov also showed that for an arbitrary $\epsilon > 0$ $N_\epsilon(n)/N(n) \longrightarrow 0$ as $n \longrightarrow \infty$, i.e. "almost all" Boolean functions $f \in B_n$ have a complexity nearly $C(n)$, i.e. about the maximum of complexities.

The *depth* $D(A)$ of a network A is the length (number of gates) of the longest path (from one of the inputs to one of the outputs) in A. Let $D(f) = \min D(A)$, where A ranges over all networks computing f, and let $D(n) = \max_{f \in B_n} D(f)$. It is easy to show from expression (1) that

$$D(n) \geq d \cdot n + o(n).$$

The constant d will be determined by

$$d = \log_{\max(n_i)} 2$$

where n_i ranges over all numbers of inputs of gates from G. (If $C(f) \sim \rho \cdot 2^n / n$ for $f \in B_n$ then the function f cannot be realized by a network with a number of elements smaller than $\Omega(2^n / n)$. A network with $\Omega(2^n / n)$ gates has a depth of at least $d \cdot n + o(n)$.) If we modify Lupanov's method given in [Lu58, Lu65] instead of (1) we obtain that for each Boolean function $f \in B_n$ there exists a network A^f computing them and such that

$$\max_{f \in B_n} C(A^f) \sim \frac{\rho \cdot 2^n}{n} \qquad (3)$$

and

$$\max_{f \in B_n} D(A^f) \sim d \cdot n. \qquad (4)$$

Let $f \in B_n$. A network A with nr inputs and r outputs is said to compute f simultaneously on r Boolean vectors if for any input vectors $a^1, ..., a^r$ of length n it gives $f(a^1), ..., f(a^r)$ at the corresponding outputs.

We are interested in the construction of a network A for a given Boolean function $f \in B_n$ such that A computes f simultaneously on r arbitrary Boolean vectors and that $C(A) < rC(f)$. Paul [Pa76] characterizes this as an interesting problem which he was not able to solve. But in [Uh74] we already showed that for almost all Boolean functions f such networks exist, moreover, in this paper such networks A with complexities $C(A) << rC(f)$ are constructed. The proof given in Uhlig (1974) in [Uh74] is complicated. In the present paper we give a new, very simple proof.

Lupanov has formulated in 1972 the problem of computing Boolean functions on several Boolean vectors. Earlier, in his paper [Lu65] he obtained the first results for vectors with special properties.

2.   Main Result

Our result is the following.

Theorem 2.1 *Let* $\log_2 r(n) = o(n/\log_2 n)$. *Then there are functions* $u(n)$ *and* $v(n)$ *with the following properties: For an arbitrary Boolean function* $f \in B_n$ *there exists a network* $A^{f,r(n)}$ *computing* f *simultaneously on* $r(n)$ *arbitrary Boolean vectors and satisfying expressions*

$$\max_{f \in B_n} C(A^{f,r(n)}) = u(n) \sim \frac{\rho \cdot 2^n}{n} \tag{5}$$

and

$$\max_{f \in B_n} D(A^{f,r(n)}) = v(n) \sim d \cdot n. \tag{6}$$

Proof :   Without loss of generality we can assume that for each n we have $r(n) = 2^{t_n}$ where t_n is a natural number.

The main idea of the proof is the following. For arbitrary fixed natural numbers k, t and for B_n where $n > k \cdot t$, we construct by induction on t a network $A_{f,n,k,t}$ for every $f \in B_n$ simultaneously computing f on 2^t arbitrary Boolean vectors. For this we construct all of $A_{f,n,k,1}$, then all of $A_{f,n,k,2}$ using the networks $A_{f,n,k,1}$ already constructed, then all of $A_{f,n,k,3}$ and so on.

Now we present the networks $A_{f,n,k,1}$.

For arbitrary n and k where $n > k$ and for an arbitrary Boolean function $f = f(x_1, ..., x_n) \in B_n$, we take all its subfunctions depending on the variables $x_1, ..., x_{n-k}$.

If $a_{n-k+1} \cdots a_n$ is the binary representation of i then let us denote the function $f(x_1, ..., x_{n-k}, a_{n-k+1}, ..., a_n)$ by f_i.

Let $g_0 = f_0$, $g_\ell = f_{\ell-1} \oplus f_\ell$ for $1 \le \ell \le 2^k - 1$ ($\oplus$ is the sum modulo 2) and $g_{2^k} = f_{2^k-1}$. Then $f_i = g_0 \oplus ... \oplus g_i = g_{i+1} \oplus ... \oplus g_{2^k}$.

Let $b_{n-k+1} \cdots b_n$ be the binary representation of j and let us assume for the moment that $i \le j$. Then we represent

$$f(a_1, ..., a_n) = g_0(a_1, ..., a_{n-k}) \oplus ... \oplus g_i(a_1, ..., a_{n-k}), \tag{7}$$
$$f(b_1, ..., b_n) = g_{j+1}(b_1, ..., b_{n-k}) \oplus ... \oplus g_{2^k}(b_1, ..., b_{n-k}). \tag{8}$$

Otherwise, if $i > j$, then we represent

$$f(b_1, ..., b_n) = g_0(b_1, ..., b_{n-k}) \oplus ... \oplus g_j(b_1, ..., b_{n-k}), \tag{9}$$
$$f(a_1, ..., a_n) = g_{i+1}(a_1, ..., a_{n-k}) \oplus ... \oplus g_{2^k}(a_1, ..., a_{n-k}). \tag{10}$$

Network $A_{f,n,k,1}$ has input nodes $x_1, ..., x_n$ for the Boolean vector $(a_1, ..., a_n)$ and input nodes $y_1, ..., y_n$ for the Boolean vector $(b_1, ..., b_n)$. It contains a subnetwork G_ℓ (see Figure 1) for each function g_ℓ, $\ell = 0, 1, ..., 2^k$ realizing them. Since expressions (3) and (4) can be written as

$$\max_{f \in B_n} C(A^f) = \frac{\rho \cdot 2^n}{n}(1 + \beta(n)),$$
$$\max_{f \in B_n} D(A^f) = d \cdot n(1 + \gamma(n))$$

where $\beta(n) \longrightarrow 0$ and $\gamma(n) \longrightarrow 0$ as $n \longrightarrow \infty$ and since $g_\ell \in B_{n-k}$ we obtain

$$C(G_\ell) \le \frac{\rho \cdot 2^{n-k}}{n-k}(1 + \beta(n-k)), \tag{11}$$
$$D(G_\ell) \le d \cdot (n-k)(1 + \gamma(n-k)). \tag{12}$$

Network $A_{f,n,k,1}$ also contains a subnetwork $B_{n,k}$ with the inputs $x_1, ..., x_n$ and $y_1, ..., y_n$ for the vectors $(a_1, ..., a_n)$ and $(b_1, ..., b_n)$ and with $(2^k + 1)(n - k)$ outputs. Exactly $n - k$ outputs of $B_{n,k}$ are connected to the inputs of G_ℓ. They put out the Boolean vector $(a_1, ..., a_{n-k})$ if either $\ell \le i \le j$ or $\ell > i > j$ and the Boolean vector $(b_1, ..., b_{n-k})$ otherwise.

Network $B_{n,k}$ can be constructed in such a way that

$$C(B_{n,k}) \le \left(2^k + 1\right) c_1 \cdot n, \tag{13}$$
$$D(B_{n,k}) \le c_2 \cdot k \tag{14}$$

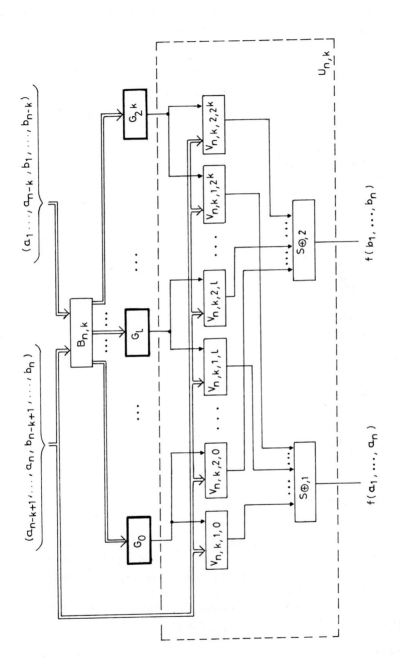

where c_1 and c_2 are sufficiently large constants. (In reality, the bound $c_2 \cdot k$ can be improved to $c_2 \cdot \log_2 k$, but the bound $c_2 \cdot k$ is sufficient for our purpose.)

Network $A_{f,n,k,1}$ also contains a subnetwork $U_{n,k}$ with $2k$ inputs for vectors $(a_{n-k+1}, ..., a_n)$ and $(b_{n-k+1}, ..., b_n)$ and $2^k + 1$ further inputs for the output signals of G_ℓ, $\ell = 0, ..., 2^k$. It has a first output for $f(a_1, ..., a_n)$ and a second output for $f(b_1, ..., b_n)$.

Network $U_{n,k}$, in its turn, consists of two networks $S_{\oplus,j}$, $j = 1, 2$, both computing the $\oplus$-sums (sum modulo two) of $2^k + 1$ variables. The outputs of $S_{\oplus,1}$ and $S_{\oplus,2}$ are the first and second outputs of the network $A_{f,n,k,1}$ respectively.

Let $c_3, c_4, ...$ be sufficiently large constants, like c_1 and c_2. We construct $S_{\oplus,j}$, $j = 1, 2$, in such a way that

$$
\begin{aligned}
C(S_{\oplus,j}) &\leq c_3 \cdot 2^k, \\
D(S_{\oplus,j}) &\leq c_4 \cdot k.
\end{aligned}
$$

Obviously this is possible.

Network $U_{n,k}$ also contains subnetworks $V_{n,k,j,\ell}$, $j = 1, 2$, $\ell = 0, 1, ..., 2^k$.

Network $V_{n,k,j,\ell}$ computes a Boolean function $f_{n,k,j,\ell}$ depending on $2k + 1$ variables and defined as follows:

$$
\begin{aligned}
&f_{n,k,1,\ell}(a_{n-k+1}, ..., a_n, b_{n-k+1}, ..., b_n, z) \\
&\qquad = \begin{cases} z & \text{if } \ell \leq i \leq j \text{ or } \ell > i > j, \\ 0 & \text{otherwise}, \end{cases}
\end{aligned}
$$

$$
\begin{aligned}
&f_{n,k,2,\ell}(a_{n-k+1}, ..., a_n, b_{n-k+1}, ..., b_n, z) \\
&\qquad = \begin{cases} z & \text{if } i \leq j < \ell \text{ or } i > j \geq \ell, \\ 0 & \text{otherwise}. \end{cases}
\end{aligned}
$$

Network $V_{n,k,j,\ell}$ can be constructed with complexity and depth both linear in k.

The inputs for the Boolean vector $(a_{n-k+1}, ..., a_n)$ are connected to input nodes $x_{n-k+1}, ..., x_n$ of the network $A_{f,n,k,1}$, the inputs for Boolean vector $(b_{n-k+1}, ..., b_n)$ are connected to input nodes $y_{n-k+1}, ..., y_n$, and the input z of network $V_{n,k,j,\ell}$ is connected to the output of network G_ℓ. The output of $V_{n,k,j,\ell}$ is connected to the ℓ-th input of $S_{\oplus,j}$. By (7) - (10) we obtain that the output signals of $U_{n,k}$ are $f(a_1, ..., a_n)$ and $f(b_1, ..., b_n)$.

We have

$$
\begin{aligned}
C(U_{n,k}) &\leq c_5 \cdot k 2^k, \tag{15} \\
D(U_{n,k}) &\leq c_6 \cdot k. \tag{16}
\end{aligned}
$$

By (11), (13), (15)

$$C(A_{f,n,k,1}) \leq \rho \cdot \frac{2^{n-k}}{n-k}(1 + \beta(n-k))(2^k + 1) + (2^k + 1)c_1 n + c_5 k 2^k.$$

This expression holds for every $f \in B_n$. Thus,

$$\max_{f \in B_n} C(A_{f,n,k,1}) \leq (2^k + 1)\left(\rho \cdot \frac{2^{n-k}}{n-k}(1 + \beta(n-k)) + c_7 \cdot n\right). \qquad (17)$$

Furthermore, by (12), (14), (16)

$$\max_{f \in B_n} D(A_{f,n,k,1}) \leq d \cdot (n-k)(1 + \gamma(n-k)) + c_8 \cdot k \qquad (18)$$

where

$$\begin{aligned} C(B_{n,k}) + C(U_{n,k}) &\leq (2^k + 1) \cdot c_7 \cdot n, \qquad &(19) \\ D(B_{n,k}) + D(U_{n,k}) &\leq c_8 \cdot k. \qquad &(20) \end{aligned}$$

For the induction step $t = T \longrightarrow t = T + 1$ we take the following induction hypothesis.

We assume that for $t = T$ for fixed k and for all $f \in B_n$, $n > Tk$, the networks $A_{f,n,k,T}$ have been constructed, where each of them computes the values of f on 2^T arbitrary Boolean vectors of length n and where

$$c_{n,k,T} = \max_{f \in B_n} C(A_{f,n,k,T}) \leq (2^k + 1)^T \left(\rho \cdot \frac{2^{n-Tk}}{n - Tk}(1 + \beta(n - Tk)) + 2^T c_7 \cdot n\right) \qquad (21)$$

and

$$\max_{f \in B_n} D(A_{f,n,k,T}) \leq d \cdot n(1 + \gamma(n - Tk)) + c_8 \cdot Tk. \qquad (22)$$

Then we take for the moment 2^T copies of G_ℓ, $\ell = 0, ..., 2^k$, $B_{n,k}$ and $U_{n,k}$, denoted by G_ℓ^p, $B_{n,k}^p$, $U_{n,k}^p$, $p = 1, ..., 2^T$. Connecting G_ℓ^p, $\ell = 0, ..., 2^k$, $B_{n,k}^p$ and $U_{n,k}^p$ for each p as shown earlier for the case $t = 1$, we obtain a network simultaneously computing f on 2 arbitrary Boolean vectors. Altogether we obtain 2^T such networks which we consider as one common network simultaneously computing f on $2 \cdot 2^T = 2^{T+1}$ arbitrary Boolean vectors.

Now for each ℓ we replace the networks G_ℓ^p, $p = 1, ..., 2^T$ by a network $A_{g_\ell,n-k,k,T}$ and obtain the network $A_{f,n,k,T+1}$.

We have by (19) and (21)

$$C(A_{f,n,k,T+1}) \leq (2^k + 1)c_{n-k,k,T} + 2^T(2^k + 1) \cdot c_7 \cdot n$$

$$\leq (2^k + 1)^{T+1} \left(\rho \cdot \frac{2^{n-k-Tk}}{n - k - Tk}(1 + \beta(n - k - Tk)) + 2^T c_7(n - k) \right)$$
$$+ 2^T(2^k + 1) \cdot c_7 \cdot n$$
$$\leq (2^k + 1)^{T+1} \left(\rho \cdot \frac{2^{n-(T+1)k}}{n - (T+1)k}(1 + \beta(n - (T+1)k)) + 2^{T+1} \cdot c_7 \cdot n \right)$$

from which it follows that for arbitrary k, t and $f \in B_n$ where $n > kt$

$$
\begin{aligned}
c_{n,k,t} &= \max_{f \in B_n} C(A_{f,n,k,t}) \\
&\leq (2^k + 1)^t \left(\rho \cdot \frac{2^{n-tk}}{n - tk}(1 + \beta(n - tk)) + 2^t \cdot c_7 \cdot n \right).
\end{aligned} \tag{23}
$$

Analogously, by (20) and (22) we obtain

$$\max_{f \in B_n} D(A_{f,n,k,t}) \leq d \cdot n(1 + \gamma(n - tk)) + c_8 \cdot tk. \tag{24}$$

Now we are able to show the validity of the Theorem. We have assumed that $r(n) = 2^{t_n}$ where t_n is a natural number.

Furthermore, by the condition $\log_2 r(n) = o(n/\log_2 n)$ we have

$$t_n = o(n/\log_2 n).$$

We take $k_n = \lceil \log_2 n - \frac{1}{2}\log_2\log_2 n \rceil$. Then

$$t_n k_n = o(n), \tag{25}$$

$$\frac{t_n}{2^{k_n}} \longrightarrow 0 \text{ as } n \longrightarrow \infty,$$

$$
\begin{aligned}
\left(2^{k_n} + 1 \right)^{t_n} &= 2^{k_n t_n} \left(1 + \frac{1}{2^{k_n}} \right)^{t_n} = 2^{k_n t_n} \left(1 + O\left(\frac{t_n}{2^{k_n}} \right) \right), \\
n - t_n k_n &= n + o(n), \\
2^{t_n} \cdot c_7 \cdot n &= o\left(\frac{2^{n - t_n k_n}}{n - t_n k_n} \right).
\end{aligned}
$$

Then by (23)

$$\max_{f \in B_n} C(A_{f,n,k_n,t_n}) \leq \rho \cdot \frac{2^n}{n} + o\left(\rho \cdot \frac{2^n}{n} \right)$$

and by (1)

$$\max_{f \in B_n} C(A_{f,n,k_n,t_n}) \geq C(n) \sim \rho \cdot \frac{2^n}{n}.$$

Thus,

$$\max_{f \in B_n} C(A_{f,n,k_n,t_n}) \sim \rho \cdot \frac{2^n}{n}.$$

By (24), (25) and (4)

$$\max_{f \in B_n} D(A_{f,n,k_n,t_n}) \sim d \cdot n.$$

Therefore, the networks A_{f,n,k_n,t_n} satisfy the Theorem.

References

[Lu58] O. B. Lupanov (1958) A method of circuit synthesis, J. Isv. VUZ Radiofiz. 1 pp. 120-140 (in Russian).

[Lu65] O. B. Lupanov (1965) A method of synthesis of control systems - the principle of local coding, Probl. Kibern. 14 pp. 31-110 (in Russian).

[Pa76] W. Paul (1976) Realizing Boolean functions on disjoint sets of variables, Theoretical Computer Science 2 pp. 383-396.

[Sa76] J. E. Savage (1976) The Complexity of Computing (John Wiley, New York, London, Sydney, Toronto).

[Sh49] C. Shannon (1949) The synthesis of two-terminal switching circuits, Bell Syst. Techn. J. 28 pp. 59-98.

[Uh74] D. Uhlig (1974) On the synthesis of self-correcting schemes from functional elements with a small number of reliable elements, J. Math. Notes Acad. Sci. USSR 15 pp. 558-562.

[We87] I. Wegener (1987) The Complexity of Boolean Functions (John Wiley, Chichester, New York, Brisbane, Toronto, Singapore).

Optimal Carry Save Networks

Michael S. Paterson [*]   Nicholas Pippenger [†]   Uri Zwick [‡]

Abstract

A general theory is developed for constructing the asymptotically shallowest networks and the asymptotically smallest networks (with respect to formula size) for the carry save addition of n numbers using any given basic carry save adder as a building block.

Using these optimal carry save addition networks the shallowest known multiplication circuits and the shortest formulae for the majority function (and many other symmetric Boolean functions) are obtained.

In this paper, simple basic carry save adders are described, using which multiplication circuits of depth $3.71 \log n$ (the result of which is given as the sum of two numbers) and majority formulae of size $O(n^{3.21})$ are constructed. Using more complicated basic carry save adders, not described here, these results could be further improved. Our best bounds are currently $3.57 \log n$ for depth and $O(n^{3.13})$ for formula size.

1. Introduction

The question 'How fast can we multiply?' is one of the fundamental questions in theoretical computer science. Ofman-Karatsuba [12] and Schönhage-Strassen [30] (see also [1],[9],[18],[36]) tried to answer it by minimising the

[*]Department of Computer Science, University of Warwick, Coventry, CV4 7AL, England. This author was partially supported by a Senior Fellowship from the SERC and by the ESPRIT II BRA Programme of the EC under contract # 3075 (ALCOM).

[†]Department of Computer Science, University of British Columbia, Vancouver, British Columbia, Canada V6T 1W5. This author was partially supported by an NSERC operating grant and an ASI fellowship award.

[‡]Department of Computer Science, University of Warwick, Coventry, CV4 7AL, England. This author was partially supported by a Joseph and Jeanne Nissim postdoctoral grant from Tel Aviv University and by the ESPRIT II BRA Programme of the EC under contract # 3075 (ALCOM). His present affiliation is the Department of Computer Science, Tel Aviv University, Tel Aviv 69978, Israel.

number of bit operations required, or equivalently the circuit size. A different approach was pursued by Avizienis [2], Dadda [7],[8] Ofman [20], Wallace [35] and others. They investigated the depth, rather than the size of multiplication circuits.

The main result proved by the above authors in the early 1960's was that, using a process called *Carry Save Addition*, n numbers (of linear length) could be added in depth $O(\log n)$. As a consequence, depth $O(\log n)$ circuits for multiplication and polynomial size formulae for all the symmetric Boolean functions are obtained.

They all used a component called a '3 $\rightarrow$ 2' *Carry Save Adder* $(CSA_{3\rightarrow2})$ which reduces the sum of three numbers (of arbitrary length) to the sum of only two in a (small) constant depth. It is easy to see that using $\log_{3/2} n + O(1)$ levels of such $CSA_{3\rightarrow2}$'s it is possible to reduce the sum of n numbers to the sum of only two. The resulting two numbers could be added (if required) using a *Carry Look Ahead Adder* (see [4],[13]) with additional depth $(1+o(1))\log m$ (where m is the length of the numbers).

In this paper we look more carefully at the construction of CSA-networks for carry save addition. A moment's reflection shows that any $CSA_{3\rightarrow2}$-network for the carry save addition of n numbers will have at least $\log_{3/2} n$ levels of $CSA_{3\rightarrow2}$'s. Thus, if a $CSA_{3\rightarrow2}$ is regarded as a black box to which the three inputs should be supplied simultaneously and which then returns, after a fixed delay, the two outputs simultaneously, then this naive construction is optimal. It turns out however that the best $CSA_{3\rightarrow2}$'s do not produce their two outputs simultaneously, nor do they require the three inputs to be supplied at the same time. The task of constructing networks with minimal total delay using such $CSA_{3\rightarrow2}$'s becomes much more interesting.

In general we assume that we are given a $CSA_{k\rightarrow\ell}$ whose delay characteristics are described by a *delay matrix* M. The entry m_{ij} of the matrix gives the relative delay of the i-th output with respect to the j-th input. In particular, if the k inputs to a $CSA_{k\rightarrow\ell}$ are ready at times $x_1,\ldots,x_k$, we assume that the i-th output is ready at time $y_i = \max_{1\leq j\leq k}\{m_{ij}+x_j\}$. This corresponds to taking the $\{\max,+\}$ inner product between M and x. We show how to extract from any delay matrix M the minimal constant $\delta = \delta(M)$ such that $CSA_{k\rightarrow\ell}$-networks for the carry save addition of n numbers with a total delay of $(\delta + o(1))\log n$ can be constructed using $CSA_{k\rightarrow\ell}$'s with delay matrix M. We exhibit explicit constructions achieving this optimal behaviour.

For a given implementation of a $CSA_{k\rightarrow\ell}$ using Boolean circuitry, if we define m_{ij} to be the length of the longest path from any bit of the j-th input number to any bit of the i-th output number then the above result translates immediately to a result about depth.

Several basic designs of CSA's are described in the next section. Using these designs optimally we get U_2-circuits (circuits over the unate dyadic basis $U_2 = B_2 - \{\oplus, \equiv\}$) of depth $5.42 \log n$ and B_2-circuits (circuits over the basis B_2 of all dyadic Boolean functions) of depth $3.71 \log n$ for the carry save addition of n numbers[1]. Using more complicated CSA's described in [23],[24] these results could be improved to $4.95 \log n$ and $3.57 \log n$ respectively. As a consequence, we derive circuits of depth $5.95 \log n$ and $4.57 \log n$ for the addition of n numbers (of linear length) and for the multiplication of two n bit numbers. This improves a previous result of Khrapchenko [17] and the naive estimates of Ofman and Wallace.

Multiple addition circuits (of n numbers of n bits each) are necessarily of size $\Omega(n^2)$. Our circuits are composed of $O(n)$ CSA's each of size $O(n)$ so they have this optimal size.

The Schönhage-Strassen multiplication algorithm [30] (see also [1],[18],[36]) uses the Discrete Fourier Transform (DFT) to obtain multiplication circuits of size $O(n \log n \log \log n)$. Since the computation of a DFT essentially involves multiple additions, carry save adders could be used to implement DFT's, and therefore the whole Schönhage-Strassen algorithm, in logarithmic depth (cf. [19],[36]). The implied constant factors however are much larger than those obtained here.

Another special case of multiple addition is bit counting. A counter for n bits could be obtained by carry save adding the n input bits, treating each as a number, and then adding the two output numbers. Note that the length of the two output numbers here is $O(\log n)$ so the additional depth required to add them up in this case is only $O(\log \log n)$. As a consequence we get depth $4.95 \log n$ U_2-circuits and depth $3.57 \log n$ B_2-circuits for counting. Many symmetric Boolean functions, such as majority and MOD_k for any fixed k, can be computed in depth $o(\log n)$ once the bit count is done, so the same bounds are valid for them as well. Shallower U_2-circuits for MOD_k with $k = 3, 5, 7$ were obtained by Chin [6].

An analogous theory is developed for formula size. We assume that the formula size characteristics of a $CSA_{k \to \ell}$ are described by an *occurrence matrix* N. The entry n_{ij} gives the number of appearances of the j-th input number in the formula for the i-th output number. If the k inputs to a $CSA_{k \to \ell}$ have formula sizes $x_1, \ldots, x_k$ then the i-th output number will have formula size $y_i = \sum_{j=1}^{k} n_{ij} x_j$. Note that this corresponds to the usual $\{+, \times\}$ inner

---

[1]All the logarithms in this paper are taken to base 2, unless otherwise stated.

product between N and x.

Again we show how to extract from an occurrence matrix N the minimal $\epsilon = \epsilon(N)$ such that $CSA_{k \to \ell}$-networks for the carry save addition of n numbers with formula size $n^{(\epsilon + o(1))}$ can be constructed using $CSA_{k \to \ell}$'s with occurrence matrix N, and again we describe explicit constructions with optimal behaviour.

Using the CSA designs of Section 2 optimally, we get U_2-formulae of size $O(n^{4.70})$ and B_2-formulae of size $O(n^{3.21})$ for each output bit in the carry save addition of n numbers and for many symmetric Boolean functions as before. Again, using more complicated CSA's, to be described in [24], these bounds could be improved to $O(n^{4.57})$ and $O(n^{3.13})$ respectively. These constructions improve previous results of Khrapchenko [16], Pippenger [26], Paterson [21] and Peterson [25].

Relevant results on formula size are the construction of concise B_2-formulae for MOD_k with $k = 3, 5, 7$ by Van Leijenhorst [34], the construction of monotone formulae for the majority function by Valiant [33] (see also [3]), and for general threshold functions by Boppana [3] and Friedman [11].

Depth and the logarithm of formula size are closely related. It is known for example that $\log L_{B_2}(f) \leq D_{B_2}(f) \leq 2.47 \log L_{B_2}(f)$ (in fact even that $D_{U_2}(f) \leq 2.47 \log L_{B_2}(f)$) and that $\log L_{U_2}(f) \leq D_{U_2}(f) \leq 1.81 \log L_{U_2}(f)$ (see [5],[28],[31]). These relations are insufficient for the derivation of optimal constants however, and we have to optimise separately for depth and for formula size. The known connections between B_2 and U_2, namely $D_{U_2}(f) \leq 2D_{B_2}(f)$ and $L_{U_2}(f) \leq O\left((L_{B_2}(f))^{\log_3 10}\right)$ (see [27]), are also too crude to be of any help to us.

The theories developed for depth and formula size are analogous. However some differences result from the fact that the usual $\{+, \times\}$ inner product is used in the formula size case while the less usual $\{\max, +\}$ inner product is used for depth. In particular, while the parameters that should be optimised in the formula size case are continuous, some of them are discrete in the delay case. This changes the nature of the optimisation problems involved.

A summary of the 'numerical' results obtained in this work together with the previously known results is given in Table 1.1. The right columns give the dimensions of the CSA's used. Results marked by a single star $(\star)$ are obtained using building blocks described in this paper, those with a double star $(\star\star)$ using building blocks described in [23], and those with $(\star\star\star)$ using building blocks that will appear in [24]. In three out of the four cases we improve the previously known results even using the same CSA's used by

U_2-**depth**				B_2-**depth**			
$5.12 \log n$	Khr[17]	(1978)	$7 \to 3$	$3.71 \log n$	$\star$	(1990)	$3 \to 2$
$5.07 \log n$	$\star$	(1990)	$7 \to 3$	$3.57 \log n$	$\star\star$	(1990)	$6 \to 3$
$4.95 \log n$	$\star\star$	(1990)	$11 \to 4$				

U_2-**formula size**				B_2-**formula size**			
n^5	Pip[26]	(1974)	$3 \to 2$	$n^{3.54}$	Pip[26]	(1974)	$3 \to 2$
$n^{4.62}$	Khr[16]	(1972)	$7 \to 3$	$n^{3.47}$	Pat[21]	(1978)	$3 \to 2$
$n^{4.60}$	$\star$	(1990)	$7 \to 3$	$n^{3.32}$	Pet[25]	(1978)	$3 \to 2$
$n^{4.57}$	$\star\star\star$	(1990)	$11 \to 4$	$n^{3.21}$	$\star$	(1990)	$3 \to 2$
				$n^{3.16}$	$\star\star\star$	(1990)	$7 \to 3$
				$n^{3.13}$	$\star\star\star$	(1990)	$6 \to 3$

Table 1.1. Results for multiple carry save addition.

the previous authors. In all the four cases we get further improvements by designing improved CSA's. The improvements we get are quite marginal in some of the cases. Our results, however, could only be improved by either improving our basic carry save adders or by designing circuits not based on carry save adders at all.

A preliminary report (with some inaccuracies) on the results described here appeared in [22]. Some results mentioned there are not covered here. One such is the construction of shallow circuits and short formulae for all symmetric Boolean functions. These are obtained by constructing circuits and formulae for counting, in which the less significant output bits are produced much before (or with smaller formula size than) the more significant ones.

2.   Carry Save Adders

A *k-bit full adder* (FA_k) receives k input bits and outputs $\lceil \log(k+1) \rceil$ bits representing, in binary notation, their sum. Usually k is of the form $2^\ell - 1$.

Arrays of FA's could be used to construct CSA's. A construction of a $CSA_{3 \to 2}$ using FA_3's, for example, is illustrated in Fig 2.1, where three input numbers a, b and c are carry save added to give the output pair u, v. The depth of the

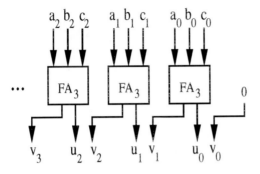

Figure 2.1. Constructing a $CSA_{3 \to 2}$ using FA_3's.

CSA obtained is equal to the depth of the FA used and is independent of the length of the numbers to be carry save added.

A B_2-implementation of an FA_3 is given in Fig. 2.2(a). The *delay matrix* of this FA_3 which describes the relative delay of each output with respect to each input, is easily seen to be $\left(\begin{smallmatrix} 1 & 2 & 2 \\ 2 & 3 & 3 \end{smallmatrix} \right)$. The delay characteristics of the $CSA_{3 \to 2}$ obtained are also described by this delay matrix. It can be checked that no other B_2-implementation of an FA_3 has a better delay matrix.

Notice that x_1 may be supplied to this FA_3 one unit of time after x_2 and x_3 are supplied, and that y_0 is obtained one unit of time before y_1. Thus, the FA_3 can be represented schematically by the 'gadget' appearing in Fig. 2.2(c).

The two formulae obtained by expanding the circuit of Fig 2.2(a) are given in Fig 2.2(b). The formula for y_1 has size 5. The variable x_1 appears only once in it while each of the variables x_2, x_3 appears twice. The *occurrence vector* of this formula is therefore $(1, 2, 2)$. The occurrence vector of the formula for y_0 is $(1, 1, 1)$. Combining these vectors we get that the *occurrence matrix* of the implementation is $\left(\begin{smallmatrix} 1 & 1 & 1 \\ 1 & 2 & 2 \end{smallmatrix} \right)$.

An alternative B_2-implementation of an FA_3 is given in Fig. 2.3. This implementation has a worse delay matrix $\left(\begin{smallmatrix} 1 & 2 & 2 \\ 3 & 3 & 3 \end{smallmatrix} \right)$ but a better occurrence matrix $\left(\begin{smallmatrix} 1 & 1 & 1 \\ 1 & 1 & 3 \end{smallmatrix} \right)$. It can be checked that no other B_2-implementation has a better occurrence matrix.

Both the implementations of Fig. 2.2 and Fig. 2.3 are also minimal with respect to circuit size (for relevant results concerning circuit size see [29],[32] and [37]). This, however, will not happen in general. Since we are not concerned here with the size of the circuits (which will always be $O(n^2)$) we

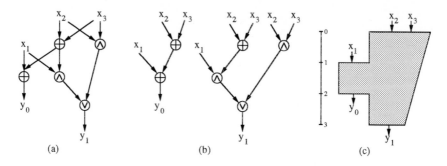

Figure 2.2. An optimal depth implementation of a B_2-FA_3.

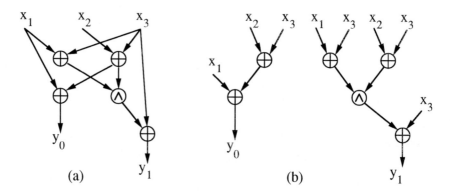

Figure 2.3. An optimal formula size implementation of a B_2-FA_3.

can think of an implementation of an FA as a set of formulae, one for each output bit. This also stresses the fact that the structure of each one of these formulae could be optimised separately.

A U_2-implementation of an FA_3 is given in Fig. 2.4. It has delay matrix $\begin{pmatrix} 2 & 4 & 4 \\ 2 & 3 & 3 \end{pmatrix}$ and occurrence matrix $\begin{pmatrix} 2 & 4 & 4 \\ 1 & 2 & 2 \end{pmatrix}$. It can be checked that both these are optimal. Note that this time it is not clear how to schedule the inputs to this unit. If x_1 is supplied two units of time after x_2, x_3, then delays are introduced in the circuitry for y_1; if x_1 is supplied one unit of time after x_2, x_3, then delays will be introduced in the y_0 circuitry. These alternatives give rise to the two gadgets shown on the right in Fig. 2.5. This is a simple example of *non-modularity*.

Using the implementation of Fig. 2.2 we get depth $3.71 \log n$ B_2-circuits for carry save addition, and using that of Fig. 2.3 we can get $O(n^{3.21})$ B_2-formulae for each output bit of carry save addition. With the implementation of Fig. 2.4 we get depth $5.42 \log n$ U_2-circuits and $O(n^{4.70})$ formulae for carry save ad-

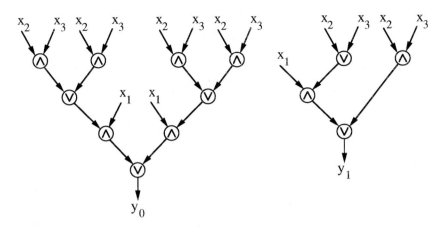

Figure 2.4. An optimal U_2-implementation of an FA_3

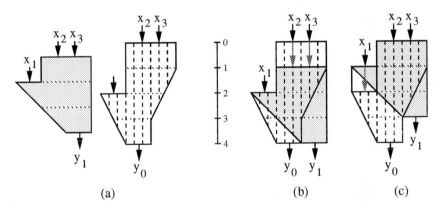

Figure 2.5. The time characteristics of the U_2-FA_3.

dition. These are the best results possible using $CSA_{3 \to 2}$'s based on FA_3's. The better results stated in the previous section are obtained using more complicated building blocks.

The construction of the building blocks that are used to get our best results is technically involved. Since we want to concentrate on the general theory we will not describe their construction here. These constructions will appear in [23],[24]. We will just point out here that CSA's could be built using any *bit adder*, not necessarily using full adders. Our currently best U_2 results, for example, are obtained using a bit adder that adds 7 bits with weight 1 together with 4 bits of weight 2. It is also not necessary for the result to be represented in a non-redundant form.

3.   *CSA-networks*

A *CSA* is regarded henceforth as a black box with k inputs and ℓ outputs ($\ell < k$) with the property that the sum of its outputs is always equal to the sum of its inputs. The delay and formula size characteristics of a *CSA* are described by its $\ell \times k$ delay and occurrence matrices.

We assume that all the entries in the delay matrix are positive and that all the entries in the occurrence matrix are at least 1. This corresponds to the assumption that every output depends on every input and that no computation is instantaneous.

A more general setting could allow the outputs to depend on subsets of the inputs. In such cases, $-\infty$ entries will appear in the delay matrices and 0 entries will appear in the occurrence matrices. Almost all the results of this paper could be extended to cover this more general situation. The complete proofs of these results are longer and will not be presented here. Some of them appear in [23],[24]. In the sequel, whenever a delay matrix is considered, it is therefore assumed that all its entries are positive and whenever an occurrence matrix is considered it is assumed that all its entries are at least one.

A *CSA*-network is an acyclic network composed of *CSA* units of a fixed type. An inductive argument shows that any *CSA*-network has the property that the sum of its outputs is equal to the sum of its inputs.

Given the delay and occurrence matrices, M and N respectively, of the *CSA* unit used, we can assign a delay and (formula) size to each 'wire' in the network. The inputs to a network are assigned delay 0 and size 1. If the k inputs to a *CSA* have delays $x_1, \ldots, x_k$ then the i-th output of this *CSA* will have delay $y_i = \max_{1 \leq j \leq k}\{m_{ij} + x_j\}$. We express this by $y = M \diamond x$ where $\diamond$ denotes the $\{\max, +\}$ inner product. If the k inputs to a *CSA* have sizes $x_1, \ldots, x_k$ then the i-th output of this *CSA* will have size $y_i = \sum_{j=1}^{k} n_{ij} x_j$. We abbreviate this by writing $y = Nx$ where this time the usual $\{+, \times\}$ inner product is used. The delay (respectively, size) of a network is the maximum of the delays (respectively, sizes) of its outputs.

Given a fixed $CSA_{k \to \ell}$ with delay matrix M and occurrence matrix N, our task is to use these units to construct networks with n inputs and ℓ outputs and minimal delay or formula size. We denote by $D_M(n)$ the minimal delay of such an $n \to \ell$ network and by $F_N(n)$ the minimal (formula) size of such a network. Strictly speaking, $n \to \ell$ networks exist only if $(k - \ell)$ divides $(n - \ell)$, and then use exactly $(n - \ell)/(k - \ell)$ *CSA*'s. This condition is relaxed however if we allow constant zero inputs. Such *dummy* inputs also simplify the presentation of the constructions described in Sections 8 and 9.

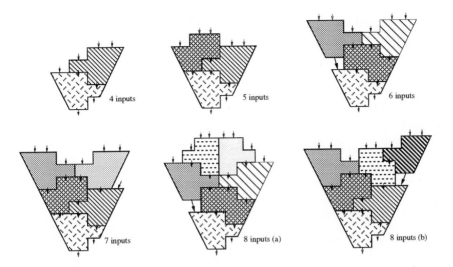

Figure 3.1. Optimal $n \to 2$ networks for $n = 4, 5, 6, 7, 8$.

The optimal networks for some small values of n, constructible using the $CSA_{3\to2}$ of Fig. 2.2, are shown in Fig. 3.1. For every fixed CSA unit there is a polynomial time algorithm which constructs for every n the set of $n \to \ell$ networks with minimal delay.

We are interested in the asymptotic behaviour of the functions $D_M(n)$ and $F_N(n)$. The next theorem states that $D_M(n)$ behaves logarithmically and $F_N(n)$ polynomially.

Theorem 3.1 *For every delay matrix M and occurrence matrix N there exist constants $\delta(M)$ and $\epsilon(N)$ such that*

(i) $D_M(n) = (\delta(M) + o(1)) \log n;$

(ii) $F_N(n) = n^{\epsilon(N)+o(1)}.$

Proof :   By collapsing first the columns then the rows of an $n \times m$ array of inputs, we see that

$$D_M(n \cdot m) \leq D_M(n) + D_M(m) + D_M(\ell^2).$$

It follows that the function $D_M(n) + D_M(\ell^2)$ behaves sub-additively (as its argument n multiplies), and thus the limit $\delta(M) = \lim_{n \to \infty} D_M(n)/\log n$ exists. By a similar argument, the limit $\epsilon(M) = \lim_{n \to \infty} \log F_N(n)/\log n$ exists. These two constants satisfy the required conditions. $\square$

Our goal in the next sections will be to determine $\delta(M)$ and $\epsilon(N)$ as functions of M and N. In the next section we give a high level overview of the results obtained in later sections.

4.   Overview of results

Given a $CSA_{k\rightarrow\ell}$ with delay matrix M, we want to construct $n \rightarrow \ell$ networks with asymptotically minimal delay. This in particular will give us the value of $\delta(M)$.

Let us assume at first that the delay matrix M is *modular*, i.e., that each pair of rows (and columns) in it differs by a constant. Alternatively, M is modular if and only if it can be represented as $b - a^T$ where $b \in \mathbb{R}^\ell$ and $a \in \mathbb{R}^k$, i.e., $m_{ij} = b_i - a_j$. The time characteristics of CSA units with modular delay matrices could be depicted schematically by strangely-shaped boxes, as was done in Figures 2.2(c) and 3.1.

Given a modular CSA, our goal is to build networks in which copies of this CSA unit are packed together tightly (in time) so that in most connections no delays are introduced. As seen from the optimal $6 \rightarrow 2$ and $8 \rightarrow 2$ networks of Figure 3.1, delays are sometimes inevitable. We will see however that we can always connect the CSA's together in such a way that the asymptotic effect of all these delays is negligible.

If $M = b - a^T$ is a modular matrix and all the entries of M are positive (this means that the first output is obtained only after the last input is supplied) then its *characteristic equation* $P_{a,b}(\lambda) = \sum \lambda^{a_j} - \sum \lambda^{b_i} = 0$ has a unique solution $\lambda = \lambda(a, b)$ in the interval $(1, \infty)$. We will prove that in d units of time, the best that we can do is to add about λ^d numbers. This immediately shows that $\delta(M) = 1/\log \lambda(a, b)$.

We next turn our attention to non-modular gadgets. We have already encountered one (see Figures 2.4 and 2.5) in our attempts to construct a U_2-FA_3. Although we cannot represent the behaviour of a non-modular gadget by a single strangely-shaped box, we can still do so for each of its outputs separately, as was done in Figure 2.5(a).

In a sense, a non-modular gadget can never be utilised fully. If a CSA with a non-modular matrix M gets its inputs at times specified by the vector x, and produces its outputs at times specified by the vector $y = M \diamond x$, then it essentially functions as a modular gadget with delay matrix $M' = y - x^T$. The matrix M' *dominates* M, i.e., every entry in M' is at least as large as the corresponding entry in M. If M is indeed non-modular, then at least one entry in M' is strictly larger than the corresponding entry in M.

Note that the collection of modular matrices dominating M forms a convex set. We shall see that it is in fact a polyhedron and we call it the *modular polyhedron over* M.

In a network composed of CSA's with a non-modular matrix M, the behaviour of each such unit can be described by a modular matrix dominating M. These modular matrices need not all be the same. The variety of forms in which a non-modular CSA could function in a network is schematically obtained by superimposing the boxes describing each of its components, allowing arbitrary shifts between them. An attempt to depict this process was made in Figures 2.5(b) and (c), where the two extremal shift positions are shown.

In the case of the gadget considered in Figure 2.5 all the delays are integral. If we assume that all the inputs are available at some integral time, say time 0, then the behaviour of each such gadget in a network will be dominated by at least one of the two modular gadgets of Figure 2.5(b),(c). We already know the best result that we could get by using each of these two gadgets separately. Can we do better by using both together?

Things could be even more complicated. If some of the delays of a non-modular gadget were non-integral then intermediate, non-extremal shifts between the different components might appear useful. If some of the delays were irrational we could have an infinite number of possible shifts to consider.

Fortunately, the situation is not so complicated. We shall see that the optimal way of using a non-modular matrix M is always to choose one modular matrix M' dominating it for which $\delta(M')$ is maximal (in certain degenerate cases there may be more than one such matrix), and then to use M' optimally on its own. Furthermore, M' could always be chosen from amongst the finitely many matrices with extremal relative shifts between the different components of M, i.e., amongst the vertices of the modular polyhedron over M.

What happens if we have at our disposal a collection of r CSA units with delay matrices $M_1, \ldots, M_r$? Again, it turns out that optimal asymptotic results could be obtained using a single CSA from this collection with a maximal value of $\delta(M_i)$ in this set.

We next consider the formula size case. Given a occurrence matrix N, we want to construct $n \to \ell$ networks in which the outputs have minimum asymptotic formula size. In the depth case we had to find the optimal relative delays that inputs to the same CSA unit should have. Analogously, we now have to find the optimal ratios between the formula sizes of inputs that are to be fed into the same CSA unit.

Given a vector $x \in \mathbb{R}^k$ and a number $p > 0$, the L_p-norm (or pseudo-norm if $p < 1$) is defined to be $\|x\|_p = (\sum |x_i|^p)^{1/p}$. The L_p-norm of an $\ell \times k$ matrix N is defined to be $\|N\|_p = \max_{x \neq 0} \frac{\|Nx\|_p}{\|x\|_p}$. The norm of a matrix gives a bound on the expansion that can occur as a result of applying N (note however that x and Nx do not lie in the same space). We need an opposite measure that we call *anti-norm* that bounds the shrinkage that can occur as a result of applying N. We define the L_p-anti-norm of N to be $\langle\!\langle N \rangle\!\rangle_p = \min_{x > 0} \frac{\|Nx\|_p}{\|x\|_p}$ where $x > 0$ means that all the components of x are positive.

If N is a occurrence matrix then the anti-norm $\langle\!\langle N \rangle\!\rangle_p$ is strictly increasing with p. There is therefore a unique number $p = p(N)$ for which $\langle\!\langle N \rangle\!\rangle_p = 1$. We will show that $\epsilon(N) = 1/p(N)$. Furthermore, the unique direction for which we have $\|Nx\|_p = \|x\|_p$ gives the formula size ratios to be used in the optimal construction for N.

5.   Lower bounds

Define the following functions :

$$\delta'(M) \;=\; \max\Big\{\delta > 0 : \begin{array}{l} \forall\, x \in \mathbb{R}^k \text{ and } y = M \diamond x \\ \sum_{j=1}^{k} 2^{x_j/\delta} \le \sum_{i=1}^{\ell} 2^{y_i/\delta} \end{array} \Big\},$$

$$\epsilon'(N) \;=\; \max\Big\{\epsilon > 0 : \begin{array}{l} \forall\, x \in (\mathbb{R}^{\ge 0})^k \\ \|x\|_{1/\epsilon} \le \|Nx\|_{1/\epsilon} \end{array} \Big\}.$$

The motivation for these definitions will become clearer in the statement and proof of Theorem 5.2 below. The following technical lemma establishes the existence of $\delta'(M)$ and $\epsilon'(N)$.

Lemma 5.1 *The maxima in the definitions of $\delta'(M)$ and $\epsilon'(N)$ exist.*

Proof :   For every $x \in \mathbb{R}^k$ and $y = M \diamond x$, let $A_x = \{\delta > 0 : \sum 2^{x_j/\delta} \le \sum 2^{y_i/\delta}\}$. It is easy to see that the set $A'_x = A_x \cup \{0\}$ is closed. Let $A = \cap_{x \in \mathbb{R}^k} A_x$ and $A' = \cap_{x \in \mathbb{R}^k} A'_x = A \cup \{0\}$. The set A', being an intersection of closed sets, is also closed.

Let $m = \max\{m_{ij}\}$ be a maximal entry in M. If we choose $x = 0$ then $y_i \le m$. If $\delta \in A_0$ then $k = \sum 2^{x_j/\delta} \le \sum 2^{y_i/\delta} \le \ell \cdot 2^{m/\delta}$ and, as a consequence, $\delta \le m/\log(k/\ell)$. Thus A_0, and therefore A', is bounded. Consequently A' is compact and has a maximum.

Let $m' = \min\{m_{ij}\}$ be a minimal entry in M. We assume that all the entries in M are positive and therefore $m' > 0$. Let $x \in \mathbb{R}^k$ and $y = M \diamond x$. Let

$x_r = \max\{x_j\}$. There exists $y_s \geq x_r + m'$, so for $\delta \leq m'/\log k$ we have $\sum 2^{x_j/\delta} \leq k \cdot 2^{x_r/\delta} \leq 2^{y_s/\delta} \leq \sum 2^{y_i/\delta}$. In particular we get that $m'/\log k \in A_x$ for every $x \in \mathbb{R}^k$, and therefore $m'/\log k \in A'$.

Thus $\max A' > 0$, and therefore $\max A = \max A'$ and their common value is $\delta'(M)$. The existence of $\epsilon'(N)$ is proved in a similar manner.   □

More direct proofs for the existence of these maxima could be derived from the arguments used in Sections 6 and 7.

In this section we show that $\delta'(M), \epsilon'(N)$ are lower bounds for $\delta(M), \epsilon(N)$. In Sections 6 and 7 we will show that they are also upper bounds, thus establishing that $\delta(M) = \delta'(M)$ and $\epsilon(N) = \epsilon'(N)$.

Theorem 5.2

(i) $D_M(n) \geq \delta'(M) \log(n/\ell)$;

(ii) $F_N(n) \geq (n/\ell)^{\epsilon'(N)}$.

Proof :   Consider an $n \to \ell$ network composed of CSA's with delay matrix M. If the inputs to a CSA in the network have delays $x_1, \ldots, x_k$ then the outputs will have delays $y_1, \ldots, y_\ell$ where $y = M \diamond x$. The definition of $\delta = \delta'(M)$ ensures that $\sum_{j=1}^{k} 2^{x_j/\delta} \leq \sum_{i=1}^{\ell} 2^{y_i/\delta}$. Using induction we get a similar relation for the inputs and outputs of the whole network. The n inputs have delay 0. If the ℓ outputs all have delay at most d then we get that $n \leq \ell \cdot 2^{d/\delta}$ or equivalently $d \geq \delta \log(n/\ell)$.

Similarly, if the CSA's in the network have occurrence matrix N we get that the sizes of the inputs and the outputs to every CSA in the network satisfy the relation $\|x\|_{1/\epsilon} \leq \|y\|_{1/\epsilon}$ where $\epsilon = \epsilon'(N)$. Using induction we get a similar relation for the whole network. The n inputs have size 1. If the ℓ outputs all have size at most f then we get that $f \geq (n/\ell)^\epsilon$.   □

The argument used in this proof is similar to the one used in a proof that a binary tree of depth ℓ can have at most 2^ℓ leaves using Kraft's inequality, that $\sum 2^{-\ell_i} \leq 1$ where the ℓ_i's are the depths of the leaves in a binary tree.

An immediate consequence is the following result.

Corollary 5.3

(i) $\delta'(M) \leq \delta(M)$;

(ii) $\epsilon'(N) \leq \epsilon(N)$.

6.   The delay problem

In this section we show how to compute $\delta'(M)$ $(= \delta(M))$ for every delay matrix M. We start by showing that, for a modular delay matrix M, we can obtain $\delta'(M)$ by solving one polynomial equation. We then show that for general, non-modular delay matrices we simply have to consider a finite number of modular matrices, compute δ' for each one of these and take the maximum. The modular matrices that we have to consider are the vertices of the modular polyhedron over M (formal definitions of all these concepts appear below).

If $x \in \mathbb{R}^k$ and $y \in \mathbb{R}^\ell$, we denote by $y - x^T$ the $\ell \times k$ matrix whose elements are $y_i - x_j$. Matrices of this form are called *modular*. We define

$$P_{x,y}(\lambda) = \sum_{j=1}^{k} \lambda^{x_j} - \sum_{i=1}^{\ell} \lambda^{y_i}$$

to be the *characteristic polynomial* of the modular delay matrix $y - x^T$.

Lemma 6.1 *If* $\max x_j < \min y_i$ *and* $\ell < k$ *then the equation* $P_{x,y}(\lambda) = 0$ *has a unique root* $\lambda(x, y)$ *in the interval* $(1, \infty)$.

Proof :   If $\lambda = 1$ then $P_{x,y}(1) = k - \ell > 0$ and if $\lambda \to \infty$ then $P_{x,y}(\lambda) \to -\infty$. Thus the equation has at least one root in the interval $(1, \infty)$.

Since a translation of x and y by the same amount leaves the roots invariant, we may assume, without loss of generality that $\max_j x_j < 0 < \min_i y_i$. Every positive contribution to $P_{x,y}(\lambda)$ is now decreasing in λ, while every negative contribution is increasing. Therefore $P_{x,y}(\lambda)$ as a whole is decreasing and the uniqueness of the root is guaranteed.   □

If M', M are two matrices and $m'_{ij} \geq m_{ij}$ for every i, j, we say that M' *dominates* M and we write $M' \geq M$.

If we replace δ in the definition of $\delta'(M)$ by $\lambda = 2^{1/\delta}$ then since $y = M \diamond x$ implies $y - x^T \geq M$, we get that $\delta'(M) = 1/\log \lambda(M)$ where

$$\lambda(M) = \min \left\{ \lambda \geq 1 : \begin{array}{l} \forall \, x \in \mathbb{R}^k, y \in \mathbb{R}^\ell, y - x^T \geq M \\ \sum_{j=1}^{k} \lambda^{x_j} \leq \sum_{i=1}^{\ell} \lambda^{y_i} \end{array} \right\}.$$

If M is a delay matrix with positive entries and $y - x^T \geq M$ then clearly $\max_j x_j < \min_i y_i$ and the uniqueness of $\lambda(x, y)$ follows. We therefore get that $\sum \lambda^{x_j} \leq \sum \lambda^{y_i}$ holds if and only if $\lambda \geq \lambda(x, y)$. We can thus state the definition of $\lambda(M)$ in the following form

$$\lambda(M) = \max\{\lambda(x, y) : y - x^T \geq M, x \in \mathbf{R}^k, y \in \mathbf{R}^\ell\}.$$

A pair (x, y) of vectors $x \in \mathbf{R}^k$ and $y \in \mathbf{R}^\ell$ satisfying $y - x^T \geq M$ will be called a *schedule* for M. If we impose the additional requirement that $x_1 = 0$ we get a one-to-one correspondence between schedules and modular matrices dominating M. This set of modular matrices dominating M will be denoted by $P(M)$ and will be called the *modular polyhedron* over M.

If $M = b - a^T$ is a modular matrix and $y - x^T \geq M$ then $y_i \geq \max_j\{b_i - a_j + x_j\} = b_i + c$ where $c = \max_j\{x_j - a_j\}$. If we define $x'_j = c + a_j$ then we still have $y - x'^T \geq M$, although $x' \geq x$ and therefore $\lambda(x', y) \geq \lambda(x, y)$. Since translating both x and y by the same amount leaves the roots invariant, we may assume that $c = 0$. So $x' = a, y = b$, and we have proved the following theorem.

Theorem 6.2 *If M is modular, $M = b - a^T$ say, then $\lambda(M) = \lambda(a, b)$.*

As an immediate consequence of this theorem we get that

$$\lambda(M) = \max\{\lambda(M') : M' \in P(M)\}.$$

The set $P(M)$ is defined using a finite set of linear inequalities and it is therefore a polyhedron. As mentioned before, we can identify a point $M' \in P(M)$ with the unique schedule (x, y) that satisfies $x_1 = 0$ and $y - x^T = M'$.

For every schedule (x, y), define a bipartite graph $\Gamma(x, y)$ in which the elements of x and y are the nodes and in which x_j and y_i are connected by an edge if and only if $y_i - x_j = m_{ij}$. A vertex of a polyhedron is an extremal point of it, that is, a point which is not a convex combination of any other two points in the polyhedron. It is easy to check that (x, y) is a vertex of the polyhedron $P(M)$ if and only if the graph $\Gamma(x, y)$ is connected. The polyhedron $P(M)$ has only a finite number of vertices. We denote this finite set of vertices by $P^*(M)$. Our aim is to prove that the maximum in the previous relation for $\lambda(M)$ is attained at some vertex of $P^*(M)$.

Suppose that $(x, y) \in P(M)$ is not a vertex of $P(M)$, so the graph $\Gamma(x, y)$ is disconnected. We will show that there exists a schedule (x', y') such that $\Gamma(x', y')$ is connected and $\lambda(x', y') \geq \lambda(x, y)$. Suppose that A is the set of variables in the nonempty connected component of $\Gamma(x, y)$ containing x_1. Let

B denote the complementary set of variables. We can split up the definition of $P_{x,y}(\lambda)$ in the following way

$$P_{x,y}(\lambda) = \underbrace{\left(\sum_{x_j \in A} \lambda^{x_j} - \sum_{y_i \in A} \lambda^{y_i} \right)}_{P_A(\lambda)} + \underbrace{\left(\sum_{x_j \in B} \lambda^{x_j} - \sum_{y_i \in B} \lambda^{y_i} \right)}_{P_B(\lambda)}.$$

Let $\lambda = \lambda(x,y)$, so that $P_{x,y}(\lambda) = 0$. If $P_B(\lambda)$ is negative, then decreasing the variables of B by a common constant increases $P_B(\lambda)$. While if $P_B(\lambda)$ is non-negative, then increasing these variables by a common constant does not decrease $P_B(\lambda)$. In either case the variables of B can be shifted in the appropriate direction until one more of the constraints $y_i - x_j \geq m_{ij}$ is satisfied with equality. The result is a schedule (x', y') for which $\Gamma(x', y')$ has all the edges of $\Gamma(x, y)$ together with at least one edge between A and B. Furthermore $P_{x',y'}(\lambda) \geq 0$ for $\lambda = \lambda(x, y)$, so that $\lambda(x', y') \geq \lambda(x, y)$.

By repeating this procedure we arrive at a schedule for which the graph is connected (so that the schedule corresponds to a vertex of $P(M)$), without ever decreasing λ.

We have thus proved

Theorem 6.3 *For any delay matrix M with positive entries*

$$\lambda(M) = \max\{\lambda(M') : M' \in P^*(M)\}.$$

As a simple example we may consider the delay matrix $\begin{pmatrix} 2 & 4 & 4 \\ 2 & 3 & 3 \end{pmatrix}$ corresponding to the *CSA* unit of Figure 2.5. The two gadgets in Fig. 2.5(b) and (c) are the vertices of the modular polyhedron. The gadget in (c) turns out to be the better and thus $\delta' \begin{pmatrix} 2 & 4 & 4 \\ 2 & 3 & 3 \end{pmatrix} = \delta' \begin{pmatrix} 3 & 4 & 4 \\ 2 & 3 & 3 \end{pmatrix} \simeq 5.42$. More details on this example may be found in Section 10.

As mentioned earlier, the results of this section could be generalised to cover the case in which a finite set of basic gadgets, each with an associated delay matrix, is given to us. In order to get optimal asymptotic performance it is always enough to use only one of the available gadgets.

7.   The formula size problem

Our aim in this section is to prove that if N is an occurrence matrix and $\epsilon = \epsilon'(N)$ then there exists a unique direction $x \in (\mathbb{R}^{>0})^k$ (where $\mathbb{R}^{>0} = (0, \infty)$) for which $\|x\|_{1/\epsilon} = \|Nx\|_{1/\epsilon}$. The existence of such a positive direction will be needed in Section 9 where constructions with formula size achieving the lower bound from Section 5 are obtained. This special direction gives the optimal ratios between the sizes of inputs that are to be fed into a CSA unit with occurrence matrix N.

We will need the following lemma.

Lemma 7.1 *If N is an occurrence matrix then $\epsilon'(N) > 1$.*

Proof : It is clear that if $N' \geq N$ then $\epsilon'(N') \geq \epsilon'(N)$. The $k \times \ell$ occurrence matrix **1** all of whose entries are 1 is dominated by every other $k \times \ell$ occurrence matrix. A direct computation shows that $\epsilon'(\mathbf{1}) = \log_{k/\ell} k > 1$. □

We are now ready to prove

Theorem 7.2 *If N is an occurrence matrix and $\epsilon = \epsilon'(N)$ then there exists a unique direction $x \in (\mathbb{R}^{>0})^k$ for which $\|x\|_{1/\epsilon} = \|Nx\|_{1/\epsilon}$.*

Proof : Consider the function

$$f(x) = \left(\|Nx\|_{1/\epsilon}\right)^{1/\epsilon} = \sum_i \left(\sum_j n_{ij}x_j\right)^{1/\epsilon}$$

over the set $B = \{x \in (\mathbb{R}^{\geq 0})^k : \|x\|_{1/\epsilon} = 1\}$. The function f is continuous and the set B is compact so the function f has a minimal point x^* in B. If $f(x^*) < 1$ we get a contradiction to the requirement that $\|x\|_{1/\epsilon} \leq \|Nx\|_{1/\epsilon}$ for every $x \in (\mathbb{R}^{\geq 0})^k$. If $f(x^*) > 1$ we get a contradiction to the maximality of ϵ. Therefore $f(x^*) = 1$. This establishes the existence of a direction in $(\mathbb{R}^{\geq 0})^k$ with the required equality.

In order to prove the uniqueness of x^*, define $x'_j = x_j^{1/\epsilon}$ and $n'_{ij} = n_{ij}^{1/\epsilon}$. We now have

$$f(x) = f'(x') = \sum_i \left(\sum_j \left(n'_{ij}x'_j\right)^\epsilon\right)^{1/\epsilon}$$

and the domain is $B' = \{x \in (\mathbb{R}^{\geq 0})^k : \sum_j x'_j = 1\}$. A minimum point of f on B corresponds to a minimum point of f' on B'. Lemma 7.1 tells us that $\epsilon > 1$. It is well known that for $\epsilon > 1$ the L_ϵ norm, $\|x\|_\epsilon = \left(\sum_j |x_j|^\epsilon\right)^{1/\epsilon}$, is strictly convex, that is, $\|\alpha x + (1-\alpha)y\|_\epsilon < \alpha\|x\|_\epsilon + (1-\alpha)\|y\|_\epsilon$ for $0 < \alpha < 1$, provided that x is not a scalar multiple of y. For every i the function $f'_i(x') = \left(\sum_j \left(n'_{ij}x'_j\right)^\epsilon\right)^{1/\epsilon}$ is obtained from $\|x\|_\epsilon$ by scaling and it is therefore also strictly convex. The function $f'(x')$ is obtained by summing strictly convex functions and it is therefore strictly convex also. The set B' is convex and therefore f' has a unique minimum point on it.

Finally, we prove that the minimum point x^* lies in the interior of B' so that none of its co-ordinates is 0. Let $f_i''(x') = (f_i'(x'))^\epsilon = \sum_j (n_{ij}' x_j')^\epsilon$. Suppose, on the contrary, that one of the co-ordinates of x^* were zero. We know of course that at least one of the co-ordinates of x^* is non-zero. Without loss of generality assume that $x_1^* = 0, x_2^* > 0$. It is easy to check that the function $(n_{i1}\Delta)^\epsilon + (n_{i2}(x_2 - \Delta))^\epsilon$ has a negative derivative at $\Delta = 0$. For small values of Δ we would therefore get that $f_i''(x_\Delta^*) < f_i''(x^*)$, or equivalently $f_i'(x_\Delta^*) < f_i'(x^*)$, where $x_\Delta^* = (\Delta, x_2 - \Delta, \ldots, x_n)$. Since this holds for every i we get that for sufficiently small Δ, $f'(x_\Delta^*) < f'(x^*)$ which contradicts the minimality of x^*. □

The strict convexity of the functions involved makes the numerical task of finding $\epsilon = \epsilon'(N)$ and the direction $x = x(N)$ satisfying $\|x\|_{1/\epsilon} = \|Nx\|_{1/\epsilon}$ an easy one.

8.   Depth constructions

As we saw in Section 6, for every delay matrix M there exists a modular delay matrix M' which dominates it and for which $\lambda(M) = \lambda(M')$. It is therefore enough to consider in this section only modular gadgets. A general modular gadget is shown in Figure 8.1. It has the modular delay matrix $M = b - a^T$ where we assume that $0 = a_1 \le a_2 \le \ldots a_k < b_1 \le b_2 \le \ldots \le b_\ell$ and $\ell < k$. We assume here that all the outputs are produced after all the inputs have been supplied, which is the case if all entries in the delay matrix are positive. As mentioned in Section 3, the results of this section could be extended to cover more general cases but we shall not do so here.

The characteristic equation of this gadget is

$$\sum_{j=1}^{k} \lambda^{a_j} - \sum_{i=1}^{\ell} \lambda^{b_i} = 0.$$

We know that this equation has a unique root in the interval $(1, \infty)$ which we denote by $\lambda = \lambda(a, b)$.

Our goal in this section is to prove the following theorem.

Theorem 8.1 *If $M = b - a^T$ is a modular delay matrix then $D_M(n) \le (1 + o(1))\log_\lambda n$ where $\lambda = \lambda(a, b)$.*

This immediately gives us the following.

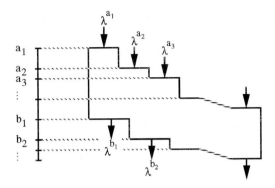

Figure 8.1. A general modular gadget.

Corollary 8.2 *For every delay matrix* M

$$\delta(M) = \delta'(M) = \max\Big\{\delta > 0 : \begin{array}{l} \forall\, x \in \mathbb{R}^k \text{ and } y = M \diamond x \\ \sum_{j=1}^{k} 2^{x_j/\delta} \le \sum_{i=1}^{\ell} 2^{y_i/\delta} \end{array}\Big\}.$$

Proof (of Theorem 8.1) :

We first consider an important special case.

Case 1. The elements of a, b are *integral*.

A *CSA* is said to lie at *level* d in a network if its inputs are supplied at times $d + a_1, \ldots, d + a_k$. Its outputs are then available at times $d + b_1, \ldots, d + b_\ell$. Since we assume here that all the delays are integral we only have to consider integral levels.

We can describe the essentials of a network by specifying the number C_d of level-d *CSA*'s used in it. The *balance* of signals generated and consumed at time d is given by

$$Bal_d = \underbrace{\sum_{j=1}^{k} C_{d-a_j}}_{\text{consumed}} - \underbrace{\sum_{i=1}^{\ell} C_{d-b_i}}_{\text{generated}}.$$

If $Bal_d > 0$ then the network accepts Bal_d new inputs at time d. If $Bal_d < 0$ then the network yields $|Bal_d|$ outputs at time d. Note that any network specified in such a way uses every signal immediately it is generated.

The choice $C_d = n\lambda^{-d}$ would yield $Bal_d \equiv 0$ and therefore represents an equilibrium. At each time the number of signals produced is equal to the number of signals consumed. The number of signals processed at time d is $\alpha \cdot n\lambda^{-d}$ where $\alpha = \sum \lambda^{a_j} = \sum \lambda^{b_i}$. The number of signals processed at time 0

is therefore $\Omega(n)$ and the number of signals processed at time $\log_\lambda n$ is $O(1)$. To turn this into an actual construction we have to tackle two problems. First, the above process is infinite, it begins at time $-\infty$ and never ends. Secondly, the number of gadgets that should be used at each time unit is generally not integral. These problems are however easy to overcome.

We choose

$$C_d = \begin{cases} \lceil n\lambda^{-d} + c \rceil & \text{if } 1 \leq n\lambda^{-d} \leq n \\ 0 & \text{otherwise} \end{cases}$$

where $c = (\ell - 1)/(k - \ell)$.

We now verify the following facts.

(i) If $0 \leq d < b_1$ then

$$Bal_d = \sum_{a_j \leq d} \lceil n\lambda^{a_j - d} + c \rceil \geq n\lambda^{-d},$$

(since $a_1 = 0$) so we may input at least n inputs at time 0, and even some more at times $1, \ldots, a_k$.

(ii) If $b_1 \leq d \leq \lfloor \log_\lambda n \rfloor$ then

$$\begin{aligned} Bal_d &\geq \sum_{j=1}^{k} \lceil n\lambda^{a_j - d} + c \rceil - \sum_{i=1}^{\ell} \lceil n\lambda^{b_i - d} + c \rceil \\ &> (n\lambda^{-d} \sum_{j=1}^{k} \lambda^{a_j}) + kc - (n\lambda^{-d} \sum_{i=1}^{\ell} \lambda^{b_i}) - \ell(c+1) \\ &= kc - \ell(c+1) = -1. \end{aligned}$$

Since Bal_d is integral we get that $Bal_d \geq 0$. In other words no outputs are produced at times less than or equal to $\lfloor \log_\lambda n \rfloor$.

(iii) If $\lfloor \log_\lambda n \rfloor < d \leq \lfloor \log_\lambda n \rfloor + b_\ell$ then

$$\begin{aligned} Bal_d &\geq -\sum_{i=1}^{\ell} C_{d-b_i} \\ &\geq -\sum_{i=1}^{\ell} \lceil n\lambda^{b_i - d} + c \rceil \\ &\geq -\lambda^{-d'} \sum_{i=1}^{\ell} \lambda^{b_i} - \ell(c+1) \quad \text{where} \quad d' = d - \lceil \log_\lambda n \rceil. \end{aligned}$$

We therefore get a total of at most $L = \frac{1}{\lambda - 1} \sum_{i=1}^{\ell} \lambda^{b_i} + \ell b_\ell(c+1)$ outputs at times less than or equal to $\lfloor \log_\lambda n \rfloor + b_\ell$. Note that L is independent of n.

(iv) Finally, if $d > \lfloor \log_\lambda n \rfloor + b_\ell$ then $Bal_d = 0$ so no more outputs are produced.

A final stage with $O(1)$ delay could now be used to reduce the L outputs obtained to only ℓ. Thus, if a, b are integral then we even have $D_M(n) \leq \log_\lambda n + O(1)$.

Note that without the assumption that $a_1 \leq a_2 \leq \ldots \leq a_k < b_1 \leq \ldots \leq b_\ell$ this construction may fail. It will not always be guaranteed that $Bal_d \geq 0$ for $1 \leq d \leq b_\ell$.

Case 2. The elements of a, b are arbitrary positive real numbers.

Consider the integer schedule

$$
\begin{aligned}
\overline{a} &= (\lfloor qa_1 \rfloor, \lfloor qa_2 \rfloor, \ldots, \lfloor qa_k \rfloor), \\
\overline{b} &= (\lceil qb_1 \rceil, \lceil qb_2 \rceil, \ldots, \lceil qb_\ell \rceil),
\end{aligned}
$$

where $q = q(n) \rightarrow \infty$ as $n \rightarrow \infty$. Note that $\overline{\lambda} = \lambda(\overline{a}, \overline{b}) \rightarrow \lambda^{1/q}$ as $q \rightarrow \infty$. The construction given in Case 1 produces $\overline{L} = O(q)$ outputs, all within time $\lfloor \log_{\overline{\lambda}} n \rfloor + \overline{b}_\ell = q(\log_\lambda n + O(1))$. Reducing the timescale in this construction by a factor of q throughout, produces a network of delay $\log_\lambda n + O(1)$ with $O(q)$ outputs. The final stage to reduce these outputs to ℓ requires a delay of $O(\log q)$. This construction satisfies the theorem provided that q is chosen so that $\log q = o(\log n)$. $\qquad \square$

9.   Formula size constructions

Let N be an occurrence matrix with entries greater than or equal to one. Choose a vector $x \in (\mathbb{R}^{>0})^k$ and define the following two vectors

$$
\begin{aligned}
a_j(x) &= \log x_j & \text{for } 1 \leq j \leq k, \\
b_i(x) &= \log \textstyle\sum_{j=1}^{k} n_{ij} x_j & \text{for } 1 \leq i \leq \ell.
\end{aligned}
$$

Associate with every CSA with occurrence matrix N, a delay matrix $M(x) = b(x) - a(x)^T$. Note that all the entries of $M(x)$ are positive.

Suppose that Γ is a network composed of CSA's with occurrence matrix N, and therefore with delay matrix $M(x)$. To every wire w in Γ we can now assign both a size $f(w)$ and a delay $d(w)$. We generalise the observation that the size of a formula is at most 2 to the power of its depth.

Lemma 9.1 *For any wire w in Γ we have $f(w) \leq 2^{d(w)}$.*

Proof :   If w is an input wire then $f(w) = 1$ and $d(w) = 0$, so the relation holds with equality. Suppose now that $u_1, \ldots, u_k$ are the inputs of some CSA in the network, that $f(u_j) \leq 2^{d(u_j)}$ for every j, and that $v_1, \ldots, v_\ell$ are the outputs of this CSA. Let $t = \max_j \{d(u_j) - a_j\}$. Then t may be regarded as the 'time' at which this CSA is activated. The delays of the inputs satisfy

$d(u_j) \le t + a_j$, while the delays of the outputs satisfy $d(v_i) = t + b_i$. The size of v_i will now be

$$f(v_i) = \sum_j n_{ij} f(u_j) \le \sum_j n_{ij} 2^{d(u_j)} \le \sum_j n_{ij} 2^{t+a_j} = 2^t \sum_j n_{ij} x_j = 2^{t+b_i} = 2^{d(v_i)}$$

and the claim of the lemma follows by induction. $\square$

Now let $x \in (\mathbb{R}^{>0})^k$ be a vector that satisfies $\|x\|_{1/\epsilon} = \|Nx\|_{1/\epsilon}$ where $\epsilon = \epsilon'(N)$. The existence of such a vector was proved in Section 7. The constructions of the previous section give $n \to \ell$ networks composed of CSA's with delay matrix $M = M(x)$ (and occurrence matrix N) with depth $(\delta(M) + o(1)) \log n$. The previous lemma shows that these networks have formula size $n^{\delta(M)+o(1)}$.

Finally, we have $\delta(M) = \delta'(M) = \epsilon'(N)$ since M is a modular matrix, and therefore $\delta'(M) = \delta$ satisfies $\sum 2^{a_j/\delta} = \sum 2^{b_i/\delta}$, or equivalently $\sum x_j^{1/\delta} = \sum (\sum_j n_{ij} x_j)^{1/\delta}$. We thus get

Theorem 9.2 *For every occurrence matrix N we have $F_N(n) \le n^{\epsilon'(N)+o(1)}$.*

An immediate consequence is the following result.

Corollary 9.3 *For every occurrence matrix N*

$$\epsilon(N) = \epsilon'(N) = \max\{\epsilon > 0 : \forall\, x \in (\mathbb{R}^{\ge 0})^k, \ \|x\|_{1/\epsilon} \le \|Nx\|_{1/\epsilon}\}.$$

10.   Numerical examples

The delay matrix of the FA_3 described in Fig. 2.2(a) is the modular matrix $M = (2\ 3)^T - (1\ 0\ 0)$. Therefore $\lambda(FA_3)$ is the unique positive root of the cubic equation $\lambda^3 + \lambda^2 - \lambda - 2 = 0$. We can verify that $\lambda \simeq 1.2056$ and that $\log_\lambda n \simeq 3.71 \log n$.

The delay matrix of the FA_3 described in Fig. 2.4 is the non-modular matrix $\begin{pmatrix} 2 & 4 & 4 \\ 2 & 3 & 3 \end{pmatrix}$. The two vertices of the modular polyhedron are $\begin{pmatrix} 2 & 4 & 4 \\ 2 & 4 & 4 \end{pmatrix} = (4\ 4)^T - (2\ 0\ 0)$ and $\begin{pmatrix} 3 & 4 & 4 \\ 2 & 3 & 3 \end{pmatrix} = (4\ 3)^T - (1\ 0\ 0)$ (see Fig. 2.5). The characteristic equation of the first vertex is $2\lambda^4 - \lambda^2 - 2 = 0$ and its root is $\lambda_1 = \sqrt{\frac{1+\sqrt{17}}{2}} \simeq 1.1317$. The characteristic equation of the second vertex is $\lambda^4 + \lambda^3 - \lambda - 2 = 0$ and its root is $\lambda_2 \simeq 1.1365$. The second possibility is clearly better and the circuits that we could build would have depth $\log_{\lambda_2} n \simeq 5.42 \log n$.

Khrapchenko [17] designed a U_2-FA_7 with delay matrix $\begin{pmatrix} 4 & 6 & 6 & 6 & 6 & 6 \\ 5 & 6 & 6 & 7 & 7 & 7 & 7 \\ 5 & 6 & 6 & 6 & 6 & 6 \end{pmatrix}$.

Using ad hoc methods he was able to construct with it networks of depth $5.12 \log n$. The delay matrix of Khrapchenko's FA_7 is non-modular. The optimal vertex in the modular polyhedron of this matrix is $\begin{pmatrix} 5 & 6 & 6 & 6 & 6 & 6 & 6 \\ 6 & 7 & 7 & 7 & 7 & 7 & 7 \\ 5 & 6 & 6 & 6 & 6 & 6 & 6 \end{pmatrix}$

and therefore $\lambda(FA_7)$ is the unique positive root of the equation $\lambda^7 + 2\lambda^6 - \lambda - 6 = 0$. We find that $\lambda \simeq 1.1465$ and that $\log_\lambda n \simeq 5.07 \log n$. We can thus improve Khrapchenko's construction even using his own gadget. We can further reduce the depth of the U_2-circuits to $4.95 \log n$ using the novel design of $CSA_{11 \to 4}$ described in [23],[24].

The size of the optimal formulae for multiple carry save addition that can be obtained using CSA's based on the FA_3 described in Fig. 2.2 is $n^{\epsilon+o(1)}$ where

$$\epsilon = \max \left\{ \begin{array}{ll} 1 & \forall x_1, x_2, x_3 \geq 0 \\ \dfrac{}{p} : & x_1^p + x_2^p + x_3^p \leq (x_1 + x_2 + x_3)^p + (x_1 + x_2 + 3x_3)^p \end{array} \right\}$$

A numerical solution gives $\epsilon \simeq 3.2058$ and equality is achieved when $x_1 = x_2 = 1$, $x_3 \simeq 0.3926$. This yields formulae of size $O(n^{3.21})$. As mentioned before we can get better results using more complicated CSA's.

11. Concluding remarks

Many related open problems still remain. The hardest of them all is probably to determine the exact depth and formula size of multiplication, multiple addition and multiple carry save addition. In this paper upper bounds and a restricted form of lower bounds on these complexities were obtained. Although the upper bounds obtained may be close to the real values, they can probably be further improved by devising better basic building blocks.

The best non-restricted lower bounds currently known for the formula size of the majority function are $\Omega(n \log n)$ in the B_2 case, due to Fischer, Meyer and Paterson [10], and $\Omega(n^2)$ in the U_2 case due to Khrapchenko [14],[15] (see also [38]). From these, $\Omega(m \cdot n \log n)$ and $\Omega(m \cdot n^2)$ lower bounds on the B_2 and U_2 formula sizes of one of the most significant output bits in the result of a multiple addition of n m-bit numbers are easily obtained. Note that such a bit, regarded as a function of the $n \times m$ input matrix, contains as a subfunction the majority function of nearly every column in the input matrix.

For multiple carry save addition a trivial $\Omega(n \log n)$ lower bound can be obtained by noticing that each set of two consecutive output pairs must depend on almost all the $n \log n$ input bits within $\log n$ positions to the right.

It can be checked that if $2^{n+1} \equiv 1 \pmod{k}$, then the $(n+1)$-st output bit of $x \cdot \left\lfloor \frac{2^{n+1}}{k} \right\rfloor$ is 1 if and only if $x > 0$ and $x \bmod k = 0$, or if $x \bmod k > \left\lceil \frac{k}{2} \right\rceil$. Using the methods of Fischer, Meyer and Paterson and of Khrapchenko, referred to earlier, we can therefore get an $\Omega(n \log n)$ B_2 lower bound and an $\Omega(n^2)$ U_2 lower bound for the formula size of multiplication.

In the U_2 case, the $\Omega(n^2)$ formula size lower bounds on these problems imply non-trivial $2 \log n - O(1)$ lower bounds for depth.

As can be seen, there are still large gaps between the upper and lower bounds and narrowing these gaps, especially by improving the lower bounds, is a challenging problem.

References

[1] Aho A.V., Hopcroft J.E., Ullman J.D., *The Design and Analysis of Computer Algorithms. Addison-Wesley, 1974.*

[2] Avizienis A., *Signed-digit number representation for fast parallel arithmetic. IEEE Trans. Elec. Comp. Vol. EC10 (1961), pp. 389-400.*

[3] Boppana R.B., *Amplification of probabilistic Boolean formulas. Advances in Computing Research, Vol. 5 on Randomness and Computation, Editor S. Micali, JAI Press, Greenwich, Conn., pp. 27-45.*

[4] Brent R., *On the addition of binary numbers. IEEE Trans. on Comp., Vol. C-19 (1970), pp. 758-759.*

[5] Brent R., Kuck D.J., Maruyama K., *The parallel evaluation of arithmetic expressions without division. IEEE Trans. on Comp., Vol. C-22 (1973), pp. 532-534.*

[6] Chin A., *On the depth complexity of the counting functions. Information Processing Letters 35 (1990) 325-328.*

[7] Dadda L., *Some schemes for parallel multipliers. Alta Frequenza, Vol. 34 (1965), pp. 343-356.*

[8] Dadda L., *On parallel digital multipliers. Alta Frequenza, Vol. 45 (1976), pp. 574-580.*

[9] Dunne P.E., *The Complexity of Boolean Networks. Academic Press, 1988.*

[10] Fischer M.J., Meyer A.R., Paterson M.S., $\Omega(n \log n)$ *lower bounds on length of Boolean formulas. SIAM J. Comput., Vol. 11 (1982), pp. 416-427.*

[11] Friedman J. *Constructing $O(n \log n)$ size monotone formulae for the k-th threshold function on n Boolean variables.* SIAM J. Comput., Vol. 15 (1986), pp. 641-654.

[12] Karatsuba A., Ofman Y., *Multiplication of multidigit numbers on automata.* Soviet Physics Dokl., Vol. 7 (1963), pp. 595-596.

[13] Khrapchenko V.M., *Asymptotic estimation of addition time of a parallel adder.* Problemy Kibernet., Vol. 19 (1967), pp. 107-122 (in Russian). English translation in Syst. Theory Res., Vol. 19 (1970), pp. 105-122.

[14] Khrapchenko V.M., *Complexity of the realization of a linear function in the class of π-circuits.* Mat. Zametki, Vol. 9 (1971), pp. 35-40 (in Russian). English translation in Math. Notes Acad. Sciences USSR, Vol. 9 (1971), pp. 21-23.

[15] Khrapchenko V.M., *Methods of determining lower bounds for the complexity of π-schemes.* Mat. Zametki, Vol. 10 (1972), pp. 83-92 (in Russian). Math. Notes Acad. Sciences USSR, Vol. 10 (1972), pp. 474-479 (English translation).

[16] Khrapchenko V.M., *The complexity of the realization of symmetrical functions by formulae.* Mat. Zametki Vol. 11 (1972) pp. 109-120 (in Russian). English translation in Mathematical Notes of the Academy of Sciences of the USSR Vol. 11 (1972) pp. 70-76.

[17] Khrapchenko V.M., *Some bounds for the time of multiplication.* Problemy Kibernet., Vol. 33 (1978), pp. 221-227 (in Russian).

[18] Knuth D.E., *The Art of Computer Programming, Vol. 2, Seminumerical algorithms (second edition).* Addison-Wesley.

[19] Mehlhorn K., Preparata F.P., *Area-time optimal VLSI integer multiplier with minimum computation time.* Information and Control, Vol. 58 (1983), pp. 137-156.

[20] Ofman Y., *On the algorithmic complexity of discrete functions.* Doklady Akademii Nauk SSSR, 145 pp. 48-51 (in Russian). English translation in Sov. Phys. Doklady, Vol. 7 (1963) pp. 589-591.

[21] Paterson M.S., *New bounds on formula size.* Proc. of 3rd GI conference on Theoretical Computer Science 1977, Lecture Notes in Computer Science 48, Springer-Verlag 1977, pp. 17-26.

[22] Paterson M.S., Pippenger N., Zwick U., *Faster circuits and shorter formulae for multiple addition, multiplication and symmetric Boolean functions.* Proceedings of the 31st IEEE Symposium on the Foundations of Computer Science (FOCS), St. Louis 1990.

[23] Paterson M.S., Zwick U., *Shallow multiplication circuits.* Proceedings of the 10th IEEE Symposium on Computer Arithmetic (ARITH), pp. 28-34, Grenoble 1991.

[24] Paterson M.S., Zwick U., *Shallow circuits and concise formulae for multiple addition and multiplication.* In preparation.

[25] Peterson G.L., *An upper bound on the size of formulae for symmetric Boolean functions.* TR 78-03-01, University of Washington.

[26] Pippenger N., *Short formulae for symmetric functions.* IBM report RC-5143, Yorktown Heights, NY (November 20, 1974).

[27] Pratt V.R., *The effect of basis on size of Boolean expressions.* Proc. 16th IEEE Symposium on FOCS (1975), pp. 119-121.

[28] Preparata F.P., Muller D.E., *Efficient parallel evaluation of Boolean expressions.* IEEE Trans. Comp., Vol. C-25 (1976), pp. 548-549.

[29] Red'kin N.P., *Minimal realizations of a binary adder.* Problemy Kibernet., Vol. 38 (1981), pp. 181-216,272 (in Russian).

[30] Schönhage A., Strassen V., *Schnelle Multiplikation grosser Zahlen.* Computing, Vol. 7 (1971), pp. 281-292.

[31] Spira P.M., *On time-hardware complexity tradeoffs for Boolean functions.* Proc. 4th Hawaii Int. Symp. on System Sciences (1971), 525-527.

[32] Stockmeyer L., *On the combinational complexity of certain symmetric Boolean functions.* Math. Syst. Theory, Vol. 10 (1977), pp. 323-336.

[33] Valiant L.G., *Short monotone formulae for the majority function.* Journal of Algorithms, Vol. 5 (1984), pp. 363-366.

[34] Van Leijenhorst D.C., *A note on the formula size of the "mod k" functions.* Info. Proc. Letters, Vol. 24 (1987) pp. 223-224.

[35] Wallace C.S., *A suggestion for a fast multiplier.* IEEE Trans. Electronic Comp. EC-13 (1964) pp. 14-17.

[36] Wegener I., *The Complexity of Boolean Functions*. Wiley-Teubner Series in Computer Science, 1987.

[37] Zwick U., *A 4n lower bound on the combinational complexity of certain symmetric Boolean functions over the basis of unate dyadic Boolean functions*. SIAM J. Comput., Vol. 20 (1991), pp. 499-505.

[38] Zwick U., *An extension of Khrapchenko's theorem*. Information Processing Letters 37 (1991) 215-217.